New Frontiers in Regional Science: Asian Perspectives

Volume 53

Editor-in-Chief

Yoshiro Higano, University of Tsukuba, Tsukuba, Ibaraki, Japan

More information about this series at http://www.springer.com/series/13039

Soushi Suzuki • Karima Kourtit • Peter Nijkamp
Editors

Tourism and Regional Science

New Roads

 Springer

Editors
Soushi Suzuki
Hokkai-Gakuen University
Sapporo, Japan

Karima Kourtit
The Open University of the Netherlands
Heerlen, The Netherlands

Peter Nijkamp
The Open University of the Netherlands
Heerlen, The Netherlands

ISSN 2199-5974   ISSN 2199-5982   (electronic)
New Frontiers in Regional Science: Asian Perspectives
ISBN 978-981-16-3625-7   ISBN 978-981-16-3623-3   (eBook)
https://doi.org/10.1007/978-981-16-3623-3

This Springer imprint is published by the registered company Springer Nature Singapore Pte Ltd.
The registered company address is: 152 Beach Road, #21-01/04 Gateway East, Singapore 189721, Singapore

Preface

This book provides new roads, perspectives, and a synthesis for the nexus of tourism and regional science research. Tourism has become one of the most dynamic sectors in the economy and has exhibited a structurally growing importance over the past decades. In many countries the economic significance of tourism now exceeds that of traditionally strong sectors like agriculture or transportation.

It is noteworthy that in recent times, tourism research has gained great momentum from the perspective of the leisure society; the psychological tension between hard work and a more relaxed lifestyle; and the productivity-enhancing or productivity-diminishing effects of leisure, recreation, and tourism. An abundance of new literature in the field of tourism management can also be found, for instance, in the areas of hospitality management, cultural events management, destination competitiveness policy and marketing, and transportation and logistics strategies, while much attention is also being paid to the opportunities provided by digital technology for the tourism sector. In addition, in the light of the many negative externalities of a rapidly growing tourism sector, there is also an abundant literature on the environmental and sustainability effects of tourism.

This book has the following objectives:

- To explore the interwoven connection between regional science and tourism research.
- To suggest promising pathways for innovative regional science research at the interface of tourism and space.
- To demonstrate the need for a new perspective on the tourism and regional science nexus by means of empirical studies.

This book is organised as follows.

Part I: Introduction

This part starts with a contribution by Peter Nijkamp, Karima Kourtit and Soushi Suzuki (Chap. 1), in which they introduce promising research angles that can be found in: (1) the close examination of the complex components of leisure time; (2) the income drivers and prosperity impacts of tourism; and (3) the place-based characteristics of tourism destinations. They also argue that the new role of regional science research in tourism is also to be found in the wider context of the hospitality sector and in a context-specific research framing.

Part II: Tourism, Specialisation and Competition

In the first chapter of this part, Mafalda Gómez-Vega, Juan Carlos Martín and Andrés J Picazo-Tadeo (Chap. 2) assess travel and tourism competitiveness (T&TC) of 136 world tourist destinations based on Data Envelopment Analysis (DEA) with endogenous common weights and the Technique for Order of Preference by Similarity to Ideal Solutions (TOPSIS). World tourist destinations are ranked using these approaches, and both of the resulting rankings are then compared to the ranking provided by the World Economic Forum (WEF).

Subsequently, Jorge Ridderstaat (Chap. 3) provides an analytical framework of smart tourism specialisation, integrating smart tourism, tourism specialisation, and nations' competitive advantage. This chapter contributes to the literature by analytically linking smart tourism with tourism specialisation and nations' competitiveness in a smart tourism specialisation framework.

Finally, in the last chapter of Part II, Soushi Suzuki and Peter Nijkamp (Chap. 4) provide an empirical contribution to trace key geographical factors for inbound and domestic tourism in Hokkaido Prefecture, Japan. They evaluate and compare the territorial tourism efficiency by means of Data Envelopment Analysis (DEA). It is inferred that an improvement of both access to regional airports and the frequency of international flights to regional airports may have a positive effect on inbound tourism and rural development in Hokkaido.

Part III: Tourism and Historical Heritage

In Chap. 5, Gert-Jan Burgers introduces democratisation and citizen participation in the management of cultural heritage. This chapter presents a useful case study of the so-called Ecomuseo della Via Appia, in a rural context in the southern Italian Apulia region.

Ila Maltese and Luca Zamparini (Chap. 6) analyse both the cultural and environmental features of the Navigli restoration project in Milan, together with the "non-users" benefits stemming from the slow mobility improvement, based on a Total Economic Value (TEV) approach and a Contingent Valuation Method (CVM).

Finally, Bart Neuts, Karima Kourtit and Peter Nijkamp (Chap. 7) investigate the impact of Airbnb in the city of Amsterdam on (a) the average housing value per m^2 and (b) the business growth in a tourism-centric local economy. Fixed effects Panel-Corrected Standard Errors models were used and found a significant effect of Airbnb listings on both the property value per m^2 and the touristification of neighbourhoods—modelled as the percentage of tourism-related activities in total neighbourhood business activities.

Part IV: Tourism Development, Sustainability and Resilience

In Chap. 8, Gabriela Carmen Pascariu, Bogdan-Constantin Ibănescu, Peter Nijkamp and Karima Kourtit unfold two complementary axes: (a) assessing the resilience of the tourism sector and (b) estimating the weight of tourism in the overall resilience performance of EU regions. And, several implications for regional and European policies are addressed as well, particularly related to the role of innovation and diversification in increasing the recovery speed following a disruption.

Subsequently, Jaewon Lim, Yasuhide Okuyama and DooHwan Won (Chap. 9) analyse the recent trends of inbound visitors to Busan and Fukuoka Metros and estimated the economic impact of Japanese (or Korean) visitors spending in Busan (or Fukuoka) Metro. The economic impact analyses also show the limited inter-industrial linkages of the tourism industry between Yeongnamkwon and Kyushu, the two closest cross-border neighbouring regions between Korea and Japan.

Finally, Tomaz Ponce Dentinho (Chap. 10) evaluates the impact of confinement measures associated with Covid-19 in tourism regions looking at the Azores Island through an econometric model, inspired by the Economic Base Model. Results allow to understand the impact of Covid-19 on tourism and to estimate the effects in total employment and population per island. The conclusion is that islands with more tourism suffered a major impact, but also recover faster.

We wish to express our thanks to all the contributors to this volume.

Sapporo, Japan   Soushi Suzuki
Heerlen, The Netherlands   Karima Kourtit
Heerlen, The Netherlands   Peter Nijkamp
May 2021

Contents

Part I   Introduction

1   Leisure, Tourism, and Space: A Thematic Exploration   3
    Peter Nijkamp, Karima Kourtit, and Soushi Suzuki

Part II   Tourism, Specialisation and Competition

2   Ranking World Tourism Competitiveness: A Comparison
    of Two Composite Indicators   15
    Mafalda Gómez-Vega, Juan Carlos Martín,
    and Andrés J. Picazo-Tadeo

3   Smart Tourism Specialization to Outfox the Competition:
    An Analytical Framework   37
    Jorge Ridderstaat

4   Key Geographical Factors for Inbound and Domestic
    Tourism in Hokkaido   51
    Soushi Suzuki and Peter Nijkamp

Part III   Tourism and Historical Heritage

5   Tourism, Leisure and Cultural Heritage: The Challenge
    of Participatory Planning and Design   71
    Gert-Jan Burgers

6   Analyzing Tourists' Preferences for a Restored City Waterway . . .   87
    Ila Maltese and Luca Zamparini

7   Space Invaders? The Role of Airbnb in the Touristification
    of Urban Neighbourhoods   103
    Bart Neuts, Karima Kourtit, and Peter Nijkamp

Part IV   Tourism Development, Sustainability and Resilience

8   **Tourism and Economic Resilience: Implications for Regional**
     Policies   129
     Gabriela Carmen Pascariu, Bogdan-Constantin Ibănescu,
     Peter Nijkamp, and Karima Kourtit

9   **Cross-Border Sustainable Tourism Development for**
     Busan-Fukuoka Megapolitan Cluster in Northeast Asia   149
     Jaewon Lim, Yasuhide Okuyama, and DooHwan Won

10   **Impact of Covid-19 in Tourism Regions. The Use of a Base**
     Model for the Azores   183
     Tomaz Ponce Dentinho

Part I
Introduction

Chapter 1
Leisure, Tourism, and Space: A Thematic Exploration

Peter Nijkamp, Karima Kourtit, and Soushi Suzuki

Abstract This contribution posits that the tourism sector—as part of the leisure society—deserves a more prominent position in regional science. After a general outline of trends in modern tourism, this chapter argues that promising research angles can be found in: (a) the closer examination of the complex components of leisure time, (b) the income drivers and prosperity impacts of tourism, and (c) the place-based characteristics of tourism destinations (including environmental and neighbourhood externalities). The new role of regional science research in tourism is also to be found in the wider context of the hospitality sector and in a context-specific research framing.

Keywords Tourism · Leisure · Regional science · Hospitality sector · Crowding · Context-specific policy · Big data · Digital technology

1.1 Tourism: A World of Choices

Over the past decades there has been an uninterrupted growth in tourism all over the world, where apparently the sky was the limit. Until the recent COVID-19 pandemic, the general worldwide expectation was that tourism would become the typical trademark of any highly mobile and highly developed country. Tourist destinations would then become the new power blocks in our world (cf. Glaeser et al. 2020). The only condition was that, to meet the demands of an ever-growing 'pleasure-society',

P. Nijkamp · K. Kourtit (✉)
Open Universiteit, Heerlen, The Netherlands

Alexandru Ioan Cuza University of Iasi, Iasi, Romania

University of Technology (UM6P), Ben Guerir, Morocco

Adam Mickiewicz University, Poznan, Poland

S. Suzuki
Hokkai-Gakuen University, Sapporo, Japan
e-mail: soushi-s@lst.hokkai-s-u.ac.jp

the tourist clients would need to have sufficient tourism resources available to realise their choices; these resources are: (a) leisure time, (b) financial resources, and (c) tourism space. These conditions will be briefly outlined here and more extensively discussed later.

Leisure time is defined as the time that is not used in a productive way by an individual (such as acquiring income or making profits). In the literature it is often assumed that the total time an individual has available during his/her productive age can be divided between work and leisure. Beatty and Torbert (2003) have argued that this conventional duality between work time and leisure presents a false dilemma, an argument also supported by Nijkamp (2020). Apart from an ambiguous demarcation of very different time patterns spent by individuals, any specific use of time is subject to scarcity—and the related choices—imposed by a 24-h day. In addition, alternative ways of spending scarce time may also have income, health, or well-being implications. This will be further discussed in Sect. 1.2.

A next contextual variable is income or wealth. Without these material resources, tourism would not be able to develop continuously. A noteworthy phenomenon that has emerged over the past decades is that tourism has moved from a 'happy few' activity to a 'global many' activity, with the necessary consequence of tourism overcrowding (see Matias et al. 2013). This has far-reaching implications for the frequency, nature, and duration of tourist visits, as will be argued in Sect. 1.3.

Another, and often overlooked fact is that tourism—especially after the transition from the 'happy few' to the 'global many'—is confronted with the limited space that is available for the visitors' many leisure activities. This holds not only for beaches or natural areas but also for historical sites or cultural amenities. Crowding has become one of the most visible negative externalities of tourism and may jeopardise the sustainable development of tourism destinations. Since the element of space (e.g. in a region, city, village, archaeological site, nature reserve) is at stake here, this issue is a typical phenomenon to be studied in regional science (see Sect. 1.4).

In conclusion, time, prosperity, and space are the three ingredients that together combine to achieve the successful development of sustainable tourism. The present chapter provides an overview of the three above-mentioned dimensions of sustainable development in the tourist industry (see, respectively, Sects. 1.2–1.4), followed by a contribution on the contextual conditions of the modern tourist sector (Sect. 1.5). We conclude with an outlook for the new role of regional science research in the emerging leisure society (Sect. 1.6).

1.2   Leisure and Tourism

Time allocation has been an important subject in the history of economic thinking. Economists often refer to the fundamental time study of Becker (1965), although in fact Shackle (1958) had already published earlier on the principles of time and uncertainty in economics. Time decisions may influence human welfare in various ways. For instance, through productive time use, individuals may generate more income (which can then be spent for recreational and visiting purposes) and hence generate more welfare, while a relaxed and leisurely life style may contribute to an individual's well-being (be it perhaps with a lower income and hence less possibilities for an expensive vacation). In this context, an interesting study on the relationship between tourists' well-being (or happiness) and the local provision of amenities can be found in a recent study by Bernini et al. (2020). Especially in the new economics of happiness (see, e.g., Frey 2019) leisure time is seen as vehicle for more happiness, while a more relaxed spending of time may favour innovativeness and creativity which may have a productivity-enhancing effect (see also Cui et al. 2019). In this context, Beatty and Torbert (2003) posit that leisure time may positively influence individual development during adulthood and hence may have an individual extrinsic economic value. It is thus clear that time preferences, from the perspective of the intertemporal elasticity of substitution between consumption now and in the future, are an important issue in leisure economics and thus also in tourism economics.

It may be noted here that in a recent study by Cui et al. (2019), the authors made an attempt to estimate a curvilinear relationship between GDP per capita and average leisure time (in hours) in several OECD countries. They were able to identify and estimate an inverted u-shape curve, from which the optimal level of leisure (measured in hours) could be derived. Clearly, leisure time is not identical to tourist time, but this type of trade-off analysis between leisure and productivity may highlight some of the complexities involved.

Against this background, tourism—as part of the leisure society—assumes an important place that is full of dilemmas. To enjoy free time, one needs productive time—and hence acquire income—to pay for luxury holidays. A high volume of productive time may lead to more income and wealth, but then there is no time left to enjoy leisure. In contrast, if one enjoys a great deal of leisure time, there is generally not a high opportunity to earn much money, so that the leisure time cannot be used for tourist and recreational visits. Consequently, leisure time, income and wealth creation, and tourism and recreational enjoyment form a complex interwoven phenomenon that is, in addition, also linked with quality of life, culture, and nature. For further studies on such time trade-offs and dilemmas we refer to, amongst others, Gicheva (2013), Guo (2004), London et al. (1977), Robinson and Godley (1997), and Wei (2006). It is clear that the position of modern tourism has to be considered from a broader conceptual and analytical perspective than is usually done in economics.

1.3 Prosperity and Tourism

Prosperity is one of the dominant background factors for the rise in tourism. In our welfare society, tourism has become one of the most dynamic sectors in the economy. It has exhibited a structurally growing importance over the past decades. In many countries, the economic significance of tourism now exceeds that of traditionally strong sectors like agriculture or transportation. Its strategic role in overall economic development can be illustrated by the fact that during the recent period of the economic crisis (as of 2007), the worldwide trajectory of international tourism has not shown a clear decline. It is thus pertinent that for economic and social reasons, a thorough scientific investigation of the drivers and consequences of modern tourism from a geographical or regional science perspective is warranted.

It is no surprise that in recent times, tourism research in the social sciences has gained considerable momentum, for instance, from the perspective of the leisure society; the psychological tension between a relaxed life style and hard work; and the productivity-enhancing or productivity-diminishing effects of leisure, recreation, and tourism. Tourism is a typical example of a multidisciplinary research field, as is also witnessed by the booming literature in such fields as: tourism management, nature reserves policy, hospitality management, entertainment and cultural events organisation, destination competitiveness policy and marketing, and transportation and logistics strategies. In recent years, in the literature also much attention has been paid to the efficiency- and accessibility-enhancing opportunities of digital technology for the tourism sector (e.g., electronic booking systems, social media platforms such as TripAdvisor). In addition, in the light of the emergence of the negative externalities of a rapidly growing tourism sector, there is also an existing literature on the environmental and sustainability implications of tourism, while recently a wave of literature has emerged on crowd analysis.

In the previous tourism literature, welfare or prosperity (usually proxied by income or GDP) is usually seen as the decisive factor for the growth of the tourism industry. However, the simple linear causal path from prosperity to tourism deserves some careful attention. At a macro-economic level, high incomes create many more opportunities for higher spending on tourism and leisure activities, but a smart and successful specialisation in the tourist industry leads also to more welfare for the tourist destinations.

However, there are also intermediate factors that play an indirect role in welfare creation for society or in income generation for an individual. Tourism—as a particular leisure activity—may create a more relaxed mood for the people involved, which may enhance their innovative and creative abilities. This is clearly linked to the trade-off between productive and leisure time. In an interesting modelling study by Celbis and Turkeli (2015), the authors used an extensive database from OECD countries in order to examine whether work time has a positive effect on innovative output. They find a diminishing role of work time, in the sense that, beyond a certain threshold value, the innovation-enhancing effect of work time is taken over by individual free time. Earlier research along similar lines can be found in Elsbach and Hargadon (2006) and Davis et al. (2013).

Tourism and leisure may also have another, indirect effect. Tourists are usually visitors to a destination who come from outside the area. In many cases, especially in our modern open global economy, they come from other countries. In that case, tourists bring in not only financial resources (the typical example of tourism-led growth; see Balaguer and Cantavella-Jorda 2002) but also cultural, knowledge, and entrepreneurial resources. Destination areas which are open to the acquisition of new skills or insights are exposed to a foreign invasion that brings in new and innovative ideas. Thus one may expect that tourism not only has a welfare-enhancing effect on destination areas, but also an innovation-enhancing effect, and hence an indirect welfare-enhancing impact (see for an analysis of ICT-enabled innovation in the hospitality sector Werthner and Klein 2005). A test on this 'double-dividend' effect of inbound tourists on regional innovativeness and progress—in the context of absorptive capacity theory—can be found in a recent study by Liu and Nijkamp (2019).

We may thus finally posit that in the complex tourist industry, various forces are at work: tourism-led growth, growth-led tourism, innovation-led tourism, and tourism-led innovation.

1.4   Space and Tourism

Tourism is, by definition, an economic activity with intrinsic spatial dimensions. It implies that travelling from A to B exerts impacts on local destinations. Surprisingly, tourism has only received modest attention in the history of regional science. One of the first contributions to the regional dimensions of tourism was written by the founding father of central place theory, Walter Christaller (1964), who offered some introductory thoughts on the upcoming importance of tourism in the post WW II period, write: 'There is also a branch of the economy that avoids central places and the agglomerations of industry. This is tourism. Tourism is drawn to the periphery of settlement districts as it searches for a position on the highest mountains, in the most lonely woods, along the remotest beaches'. However, it is noteworthy that Christaller's view on the happy few tourists escaping from busy agglomerations to areas of tranquillity is nowadays considered to be outdated.

Currently, we live in the age of mass tourism, where quiet beaches, unspoilt nature and serene cultural ambiance are hardly found anymore. Tourist destinations are challenged not only with the task to attract—through appropriate investments—the right tourists to the right places (see, e.g., Xu et al. 2020; Suzuki et al. 2011) but also—and increasingly more so—to manage large crowds of tourists. Paris, Venice, Barcelona, Dubrovnik, Rome, Bruges, Amsterdam, or Jerusalem, to mention only a few, are faced with a mass influx of visitors during the tourist season. Mass mobility is a feature of the New Urban World (Kourtit 2019). Relaxed tourist atmospheres are becoming rare spaces. High tourist concentrations in attractive destinations lead to a variety of negative externalities: loss of the historical character of old city centres, quality decline in natural areas, environmental decay in terms of air and water quality, overcrowded streets and squares, noise annoyance during the night, and so forth. It seems as though the positive benefits of tourism (in terms of income and employment effects) are overshadowed by the mass pressure on the scarce spaces in the city and natural areas, to the extent that residents might even feel like aliens in their own city. Tourism pressure leads to competition for the use of public space between locals and foreigners. Consequently, tourism is not only a human activity which seeks attractive places to visit but also a deliberate choice to engage in space competition elsewhere.

In the past few decades, this trend has been aggravated by the emergence of ICT, in particular the development and popularity of digital booking and information systems. Digital technology has become the major instrument for mass tourism, not only in well-known tourist places but also in remote areas. Digital technology has made the world one big tourism destination. Digital platforms such as Booking.com, TripAdvisor, or Airbnb cover a significant part of global tourist interests and travel decisions. In conclusion, tourism and local liveability are closely connected together. These forces ultimately lead to the difficult-to-manage phenomenon of the 'tragedy of the commons'.

A good illustration of the above-mentioned tension at the local level in tourist destinations is the emerging concern about the rapid rise of Airbnb facilities. Not only is the Airbnb market a rather uncontrolled but nevertheless large segment of the total hospitality and accommodation sector in many cities, but it also has serious implications for the well-being of incumbent residents (in particular, neighbourhood externalities) and for the housing prices and housing availability in densely populated city centres. In some cities (e.g. Venice, Barcelona), city centres seem to have become 'battle grounds' between tourists and locals.

Finally, one more caveat on the local or regional aspects of tourism is noteworthy. In most cases, tourism is a seasonal activity or a weekend activity. This means that the local hospitality sector in tourist destination places is continually faced with complex capacity management issues, in particular, fluctuations in local income revenues and demands for job. Clearly, the space–time dimensions of tourist visits have significant effects on the incoming benefit flows at the local level.

1.5   A Contextual Positioning of Tourism and Regional Science

It has been argued in the previous sections that the tourism sector is an important economic sector, with significant geographical dimensions. Consequently, a regional science focus on tourism requires a broader contextual positioning of this sector, as well as a more place-based focus.

In many regional development strategies, the tourism and leisure industry are seen as a convincing example of smart specialisation. This means a competitive choice to have a dedicated stimulus for a given promising sector, inter alia through the use of ICT facilities. Therefore, tourism may be regarded as a concerted and orchestrated action strategy for reaping the fruits of smart specialisation for the leisure industry (see, e.g., Li et al. 2020). Smart specialisation is understood here as a 'virtuous process of diversification through the local concentration of resources and competences in a certain number of new domains that represent possible path for transformation of productive structures' (Foray 2014). The implementation of smart specialisation calls for a guiding or leading role of regional or local governments, in close collaboration with relevant stakeholders, so as to benefit from place-based locational advantages in a competitive economy. In this context, governments may have four distinct roles as: (a) the representative of local community interests; (b) the manager of resources to deliver public goods and services; (c) a political-administrative entity with a legal competence for its territory; (d) a change-maker for new innovative prospects, or for handling drastic perturbations (see Hassink 2000; Corvers 2019).

The tourism sector has undergone a drastic change over the past few decades. Many cities and regions all over the world (e.g., Dubai, Venice, Lisbon, Rome, Vienna, Amsterdam, Barcelona) have adopted a strategy of smart tourism specialisation. This has created many benefits for these areas, but increasingly the awareness has grown that there is also a shadow side to having a dominant specialisation in the tourism industry. As mentioned above, local communities (e.g., in Barcelona or Venice) have started to question the economic benefits of mass tourism. The new contra-tourism movement calls for an intelligent response from city leaders and managers. In a policy-cycle framework, different strategic options for the urban governance of mass tourism can be distinguished: for instance, the efficient and responsive management of resources, institutional and stakeholder-based strategies to control tourism; the creation of a flexible and adaptable public control system to cope with perturbations in tourism growth effects; or a similar focus on city-specific and identity-oriented action plans so as to safeguard the historical strength of a locality. This brings us into the realm of context-specific (or contextualised) tourism policy.

The notion of context-specific local policy has gained considerable importance in the recent literature (see, e.g., Soete and Arundel 1993; Asheim and Isaksen 1997; Cooke and Morgan 1998; Corvers 2019; Tödtling and Tripple 2005; Landabaso 1997; McCann and Ortega-Argiles 2015; Suzuki et al. 2011). The idea of contextualised local or regional policy takes for granted that—in addition to generic and generally accepted factors and drivers—each locality has specific conditions (positive and negative) that are decisive for the outcome of a policy action. Such a policy is based on specific and purposive local development strategies, with both a strong involvement of the public sector and a clear orientation towards dedicated economic sectors. Then, apart having from high evolution scores on the generic strengths and weaknesses of a locality, a place also has its own indigenous and unique strong and weak constituents that shape economic (dis)advantages (reflected in the notion of 'unrivalled' regional or local resources; see Morgan and Nauwelaers 1999).

Thus, context-specific tourism policy has to be two-sided and requires: (a) consideration of, and responses to, external drivers and factors; and (b) intra-systemic governance of complex urban or regional systems. Further insight into contextualised governance requires a distinction to be made between four types of governance (viz. executive, process-oriented, regulatory, and contextual) and four types of functional roles (viz. community-based, service-oriented, auto-governance, and change agent). For further details, we refer to Corvers (2019).

In summary, the tourism sector is not a single economic sector operating in an island economy. It is shaped in a broad economic force field, with many disciplines and actors involved. It certainly deserves a more central place in regional science research.

1.6 Retrospect and Prospect

In the modern age of leisure and tourism—also called the 'epoch of experience'—there is an urgent need for a proper understanding of the complexity of the tourism, recreational, and hospitality sector. From an economic perspective, this is a very substantial sector in most economies, while it also has significant footprints in ecology and culture in destination areas. Clearly, the tourism sector is a multi-stakeholder multinational activity which calls for well-designed sustainability principles and operational management support.

Given the digital drivers of tourism, due insight into prevailing trends, tourism behaviour, marketing tools, and visitors' perceptions is needed (see, e.g., Romao et al. 2017; Kourtit et al. 2019). This calls not only for reliable statistical information (ranging from local statistical data to international Tourist Satellite Account data) or user-based micro data (e.g. based on survey questionnaires or interviews) but also for social media and user platform data ('big data'). Of course the merger of such data poses a great methodological challenge. Given the local nature of data, a combination with geo-science data will also be needed, so as to acquire a proper understanding of the spatial choice and mobility patterns of visitors, as well as of the tensions related to the common use of scarce public space in tourist areas. This would also allow for more coordinated research with transportation science, on the one hand, and cultural urbanistics research, on the other.

It is noteworthy that the widespread research on the tourism sector has not yet gained great momentum or an established position in regional science. Clearly, on regular occasions, publications on tourism behaviour and leisure behaviour (e.g., destination choice, travel mode choice, shopping motives, local impact analysis of tourism etc.) can be found in the regional science literature, but this has mainly remained an undercurrent and has not led to a booming research tradition. With a few notable exceptions, regional scientists have not achieved a high publication record in most tourism journals. A research agenda for regional scientists in the domain of tourism could at least comprise the following basic scientific activity plan:

- recognize the leisure society as an important field of research in the analysis of the space-economy;
- explore the interwoven connection between regional science research and tourism research;
- develop new and promising pathways for innovative regional science research at the interface of tourism and space;

- take the new digital data avalanche ('big data') in the hospitality sector as a great challenge for advanced regional science research, including geo-science;
- design a roadmap for a new framing of social science research as the interface of the human and the material dimensions of tourism in relation to scarce tourist space;
- demonstrate the necessity and potential for a new perspective on the tourism and regional science nexus by means of various empirical illustrations;
- outline the great economic potential of advanced digital technology for local and global tourism policy analysis.

In conclusion, the multidisciplinary character of regional science warrants a prominent position of leisure, tourism, recreation, and entertainment in the analysis of the complex space-economy.

Acknowledgements Karima Kourtit and Peter Nijkamp acknowledge the grant from the Axel och Margaret Ax:son Johnsons Stiftelse, Sweden. The authors also acknowledge the grant from the Romanian Ministry of Research and Innovation, CNCS—UEFISCDI, project number PN-III-P4-ID-PCCF-2016-0166, within the PNCDI III project ReGrowEU—Advancing ground-breaking research in regional growth and development theories, through a resilience approach: towards a convergent, balanced, and sustainable European Union (Iasi, Romania).

References

Asheim B, Isaksen A (1997) Location, agglomeration and innovation. Eur Plan Stud 5(3):299–330
Balaguer J, Cantavella-Jorda M (2002) Tourism as a long-run economic growth factor: the Spanish case. Appl Econ 34(7):877–884
Beatty J, Torbert W (2003) The false duality of work and leisure. J Manag Inq 12(3):239–252
Becker G (1965) A theory of the allocation of time. Econ J 76(299):493–517
Bernini C, Cerqua A, Pellegrini G (2020) Endogenous amenities, tourists' happiness and competitiveness. Reg Stud 54(9):1214–1225
Celbis MG, Turkeli S (2015) Does too much work hamper innovation? J Glob Policy Gov 4:97–116
Christaller W (1964) Some considerations of tourism location in Europe. Pap Reg Sci 12 (1):95–105. https://doi.org/10.1111/j.1435-5597.1964.tb01256x
Cooke P, Morgan K (1998) The associational economy. Oxford University Press, Oxford
Corvers F (2019) Designing 'context-specific' regional innovation policy. PhD dissertation, University of Leiden, Leiden
Cui D, Wei X, Wu D, Cui N, Nijkamp P (2019) Leisure time and labour productivity. Economics 13:1–24. https://doi.org/10.5018/economics-ejournal.ja2019-36
Davis LN, Davis JD, Hoisi K (2013) Leisure time invention. Organ Sci 24(5):1439–1458
Elsbach KD, Hargadon AB (2006) Enhancing creativity through mindless work: a framework of workday design. Organ Sci 17(4):470–483
Foray D (2014) From smart specialisation to smart specialisation policy. Eur J Innov Manag 17 (4):492–507
Frey B (2019) The economics of happiness. Springer, Berlin
Gicheva D (2013) Working long hours and early career outcomes in the high-end labor market. J Labour Econ 31(4):785–824
Glaeser E, Kourtit K, Nijkamp P (eds) (2020) Urban empires. Routledge, New York
Guo LF (2004) Economic analysis of leisure consumption. J Quant Techn Econ 21(4):12–21

Hassink R (2000) Regional innovation support systems. Eur Plan Stud 10(2):153–164

Kourtit K (2019) The new urban world. Shaker, Aachen

Kourtit K, Nijkamp P, Romao J (2019) Cultural heritage appraisal by visitors to global cities: the use of social media and urban analytics in urban buzz research. Sustainability 11(12):3470

Landabaso M (1997) The promotion of innovation in regional policy. Entrep Reg Dev 9:1–24

Li H, Nijkamp P, Xie X, Liu J (2020) A new livelihood sustainability index for rural revitalisation assessment. Sustainability 12:148. https://doi.org/10.3390/54120801148

Liu J, Nijkamp P (2019) Inbound tourism as a driving force for regional innovation. J Travel Res 58 (4):594–607

London M, Crandall R, Seals G (1977) The contribution of job and leisure satisfaction to quality of life. J Appl Psychol 62(3):328–334

Matias A, Nijkamp P, Sarmento H (eds) (2013) Quantitative methods in tourism economics. Springer, Berlin

McCann P, Ortega-Argiles R (2015) Modern regional innovation policy. Camb J Reg Econ Soc 17 (4):409–427

Morgan K, Nauwelaers C (eds) (1999) Regional innovation strategies. The Stationary Office, London

Nijkamp P (2020) Dolce far Niente. In: Frey BS, Schaltegger CA (eds) 21st century economics: economic ideas you should read and remember. Springer, Berlin, pp 91–93

Robinson J, Godley G (1997) The surprising ways Americans use their time. Pennsylvania State University Press, University Park

Romao J, Kourtit K, Neuts B, Nijkamp P (2017) The smart city as a common place for tourists and residents: a structural analysis of the determinants of urban attractiveness. Cities 78:67–75

Shackle GLS (1958) Time in economics. North Holland, Amsterdam

Soete L, Arundel A (eds) (1993) An integrated approach to European innovation and technology diffusion policy. Publication Office of the European Communities, Luxembourg

Suzuki S, Nijkamp P, Rietveld P (2011) Regional efficiency improvement by means of data envelopment analysis through Euclidean distance minimization including fixed input factors: an application to tourist regions in Italy. Pap Reg Sci 90:67–89

Tödtling F, Tripple M (2005) One size fits all? Res Policy 34:1203–1219

Wei X (2006) Leisure time and economic efficiency. Nankai Econ Stud 6(6):3–15

Werthner H, Klein S (2005) ICT-enabled innovation in travel and tourism. In: Walder B, Weiermair K, Perez A (eds) Innovation and product development in tourism. Erich Schmidt Verlag, Berlin, pp 71–84

Xu C, Jones C, Munday M (2020) Tourism inward investment and regional economic development effects. Reg Stud 54(9):1226–1237

Part II
Tourism, Specialisation and Competition

Chapter 2
Ranking World Tourism Competitiveness: A Comparison of Two Composite Indicators

Mafalda Gómez-Vega, Juan Carlos Martín, and Andrés J. Picazo-Tadeo

Abstract Assessing travel and tourism competitiveness (T&TC) is becoming an issue of paramount importance for stakeholders, mainly policymakers and practitioners. The availability of new databases such as the Travel and Tourism Competitiveness Report from the World Economic Forum (WEF) has boosted this interest. This chapter contributes to the current literature in this field with an assessment of the T&TC of 136 world tourist destinations. In doing so, two well-known approaches are employed: Data Envelopment Analysis (DEA) with endogenous common weights, and the Technique for Order of Preference by Similarity to Ideal Solutions (TOPSIS). World tourist destinations are ranked using these approaches, and both of the resulting rankings are then compared to the ranking provided by the WEF. Lastly, further analysis is carried out aimed at determining the main pillars and sub-pillars of T&TC, as a way to provide relevant insights to policymakers and practitioners. Our conclusion is that the two proposed methods are valid candidates for measuring the T&TC of world tourism destinations.

Keywords Travel and tourism competitiveness · Composite indicators · DEA · Endogenous weights · MCDM · TOPSIS · World Economic Forum

M. Gómez-Vega
University of Valladolid, Valladolid, Spain
e-mail: mafalda.gomez@uva.es

J. C. Martín (✉)
University of Las Palmas de Gran Canaria, Las Palmas de Gran Canaria, Spain
e-mail: jcarlos.martin@ulpgc.es

A. J. Picazo-Tadeo
University of Valencia and INTECO Research Group, Valencia, Spain
e-mail: andres.j.picazo@uv.es

© The Author(s), under exclusive licence to Springer Nature Singapore Pte Ltd. 2021
S. Suzuki et al. (eds.), *Tourism and Regional Science*, New Frontiers in Regional
Science: Asian Perspectives 53, https://doi.org/10.1007/978-981-16-3623-3_2

2.1   Introduction

The travel and tourism (T&T) industry is of paramount importance to improving the quality of life of millions of people, by driving economic growth, providing jobs, reducing poverty and fostering social tolerance (World Economic Forum 2017). The sector has seen continued expansion over recent decades—except for occasional crises, such as the reduction in tourist arrivals and income in 2009[1]—showing its strength and resistance (Oklevik et al. 2019). In fact, international tourist arrivals grew by 5% in 2018, to reach the 1.4 billion mark, 2 years ahead of the United Nations World Tourism Organization's (UNWTO) long-term forecasts. Likewise, export earnings generated by tourism amounted to US$1.7 trillion, registering an increase of 4% with respect to the previous year, thus outpacing the growth rate of the world economy (World Tourism Organization 2019b). This rapid global growth makes T&T comparable in terms of business volume with industries that have historically played a key role as driving forces of economic development (Mendola and Volo 2017). Furthermore, for the most part it has been boosted by a favourable economic environment, a growing middle class in emerging economies, technological advances, new business models, cheaper travel costs and international visa agreements (World Tourism Organization 2019a). The diversification of the tourist offer has also contributed to the growth of the T&T industry, with the emergence of new destinations that compete with traditional ones (Lee 2015). Likewise, destinations that previously lay outside the main tourist circuits, or focused on a very limited segment of demand, have begun to receive other types of visitors driven by different motivations such as an interest in heritage and natural resources, or tourists travelling for work, business and congresses (De Vita and Kyaw 2016).

The UNWTO is committed to ensuring that the rocketing growth of the T&T industry worldwide is managed in a responsible and sustainable way, while assuring the role of tourism as a key driver of social and economic development, job creation and equality. In this regard, the key findings from the UNWTO International Tourism Highlights 2019 report (World Tourism Organization 2019a) show that: (1) The Asia-Pacific region and Africa led the growth in arrivals with a 7% increase in 2018, while Asia-Pacific and Europe experienced above-average growth in tourism earnings; (2) Among the world's top 10 destinations in arrivals and receipts, France continued to lead in international tourist arrivals, while the United States remained the largest tourism earner in 2018, and Japan entered the top 10 earners ranking following 7 years of double-digit growth in international tourism receipts; (3) The top 10 tourism earners account for almost half of total tourism receipts, while the top 10 destinations in arrivals receive 40% of worldwide arrivals; (4) China

---

[1]The World Tourism Organization (UNWTO) has presented a preliminary evaluation on the impact of COVID-19 on international tourism. Given the forced closure of borders and general restrictions on travel around the world, international tourist arrivals are expected to decrease by 20–30% in 2020 compared to 2019 figures. However, the UNWTO points out that these estimates should be interpreted with caution in view of the extremely uncertain nature of the crisis.

remained the world's largest spender, with US$277 billion spent on international tourism in 2018—one-fifth of international tourism expenditure—followed by the United States; (5) Four out of five tourists visit a destination in their own region; (6) 58% of all international tourists reach their destinations by air, an increase from 46% in 2000; (7) The share of leisure travel has grown from 50% in 2000 to 56% in 2018. Leisure travel is the main purpose of tourists' visits in all world regions except the Middle East, where visiting friends and relatives, or travel for health or religious purposes predominates; and (8) The share of the world population requiring a traditional visa declined from 75% in 1980 to 53% in 2018.

The increasing importance of the T&T industry has attracted the interest of researchers, academics and practitioners, focusing on the analysis of travel and tourism competitiveness (T&TC) from multiple approaches (Sainaghi et al. 2017); among these, a burgeoning line of research is devoted to assessing T&TC by means of synthetic indicators. The purpose of travel and tourism competitiveness indices (T&TCI) is to shed light on the relative performance of given tourist destinations versus other destinations or a group of competitors. Accordingly, T&TCI provide managers and policymakers with relevant information that could help them to improve decision-making as well as the design of policies aimed at boosting competitiveness (Mendola and Volo 2017; Croes and Kubickova 2013), ultimately enhancing people's quality of life.

This chapter compares two T&TCI computed with two well-established multi-criteria-decision-making (MCDM) approaches: Data Envelopment Analysis (DEA), as proposed by Gómez-Vega and Picazo-Tadeo (2019); and the Technique for Order of Preference by Similarity to Ideal Solutions (TOPSIS). Then, the T&TCI ranking computed by these two indices is compared to that from the World Economic Forum (WEF) (World Economic Forum 2017). Our contribution is thus eminently practical, as it is aimed at helping policymakers, managers and other stakeholders to improve their decision-making. Likewise, our results also enable the identification of the most relevant pillars of T&TC, and the strengths and weaknesses of particular tourist destinations.

2.2   Background: Travel and Tourism Competitiveness

The competitiveness of a tourist destination is a multifaceted concept (Crouch 2011; Dwyer et al. 2014), which needs to be addressed taking into account its manifold dimensions and the various stakeholders involved. Croes and Kubickova (2013, p. 148) state that '. . . *a universal and precise definition* [of competitiveness] *does not exist*'. However, a widely accepted definition is that proposed by Ritchie and Crouch (2003), who define the competitiveness of a tourist destination as '. . .[the] *ability to increase tourism expenditure, to increasingly attract visitors while providing them with satisfying, memorable experiences, and to do so in a profitable way, while enhancing the well-being of destination residents and preserving the natural capital of the destination for future generations*'.

The literature in this field of research has addressed the study of T&TC employing different sources of data, including: (1) surveys administered directly to tourists (Bahar and Kozak 2007; Chen et al. 2008; Cracolici et al. 2008); (2) surveys administered to other stakeholders (Bornhorst et al. 2010; Dwyer et al. 2012); and (3) official statistics (Croes 2011; Das and Dirienzo 2010; Pulido-Fernández and Rodríguez-Díaz 2016; Zhang et al. 2011). In addition, numerous researchers have focused their efforts on building synthetic or aggregate T&TCI. A common concern in all these studies is the issue of aggregation. Gómez-Vega and Picazo-Tadeo (2019, p. 282) frame the problem as follows: '... *To weight or not to weight? And, if so, how to weight? ... are a couple of relevant questions that researchers should answer before attempting to build a composite indicator* [of T&TC]'. While it has been common practice to compute unweighted indicators—which means assigning equal weights to all the dimensions of T&TC—several papers have employed exogenous weighting schemes based on expert opinion or other *ad hoc* criteria. Conversely, other authors have used a variety of statistical and mathematical approaches to obtain the weights with which to aggregate the different dimensions of T&TC into a single composite indicator.

Mazanec and Ring (2011) use the Partial Least Squares-Path Modelling method to construct a weighting scheme for the pillars of T&TC proposed by the WEF, based on the pillars' explanatory power for the WEF T&TCI. Lan et al. (2012) employ Neural Network Analysis to establish an objective weighting scheme for those pillars. Croes and Kubickova (2013) determine the weights for the pillars based on their correlation with the T&TCI. One of the most recent contributions, however, is that of Pérez-Moreno et al. (2016), in which a multi-criteria model is used to deal with the problem of aggregation, and a solution to the issue of substitutability between pillars is proposed. To that end, the authors establish two statistical values of reference—the so-called aspiration and reservation values—and standardise the values of the pillars between these two references. They then propose the use of arithmetic means to construct three T&TCI with different degrees of compensation between pillars. Similarly, Pulido-Fernández and Rodríguez-Díaz (2016) use multi-criteria models based on a double reference point, although their approach does not allow for compensation between pillars. More recently, Gómez-Vega and Picazo-Tadeo (2019) computed a composite T&TCI using DEA (Charnes et al. 1978) and MCDM techniques (Despotis 2002, 2005). The foremost advantage of their approach is that the weights assigned to the indicators, pillars and factors of T&TC proposed by the WEF are obtained endogenously. Based on the so-called Benefit-of-the-Doubt (BoD) principle (Cherchye et al. 2007), this approach enables the calculation of the set of weights that places each tourist destination in the most favourable light when it is compared to all the other analysed destinations rated using the same weighting scheme.

2.3   Data

Official data from statistical offices and international organisations currently provide a compendium of indicators with which to study T&TC. As mentioned above, one outstanding source is the Travel & Tourism Competitiveness Report regularly published by the WEF, which provides an extensive database for a large group of countries. This database scores multiple dimensions of T&TC, using a global approach that takes into account the perspectives of industry leaders, international organisations and governments; drawing on these data it then provides a synthetic or composite T&TCI (World Economic Forum 2017).

The first Travel & Tourism Competitiveness Report was compiled by the WEF in 2007, as a way to measure the set of factors and policies that enable the sustainable development of T&T industries. The 2017 edition was developed in the context of the *World Economic Forum's Industry Programme for Aviation, Travel and Tourism*, in close collaboration with data producer partners—Bloom Consulting, Deloitte-STR Global, the International Air Transport Association (IATA), the International Union for Conservation of Nature (IUCN), the World Tourism Organization (UNWTO) and the World Travel & Tourism Council (WTTC); and industry partners—AccorHotels, Amadeus, AirAsia, Emirates, Etihad Airways, Gulfstream, HNA, Hilton Worldwide, Iberostar Group, Intercontinental Hotel Group, Jet Airways, Jumeirah, Marriott International, SAP/Concur, SpiceJet, Swiss/Deutsche Lufthansa and VISA.

The 2017 edition of the WEF Travel & Tourism Competitiveness Report provides information regarding 136 countries and 90 raw indicators, which capture diverse dimensions of competitiveness, including political, socioeconomic, structural, environmental and cultural aspects (World Economic Forum 2017). These indicators are aggregated into 14 pillars, which are further grouped into 4 factors of competitiveness; finally, these factors are aggregated into a global T&TCI (Fig. 2.1). From a methodological standpoint, at each level of aggregation, the composite indices are built as unweighted averages of the dimensions in the immediately preceding level, whether indicators, pillars or factors.

A joint analysis of the results from the 2017 edition of the WEF T&TCI and other quantitative and qualitative indicators reveal four key findings worth highlighting: (1) T&TC is improving, especially in developing countries, and particularly in the Asia-Pacific region. As the industry continues to grow, an increasing share of international visitors are coming from and travel to emerging and developing countries; (2) In an increasingly protectionist context, which is hindering global trade, the T&T industry continues building bridges rather than walls between people, as made apparent by the increasing numbers of people travelling across borders and global trends toward less restrictive visa policies; (3) With the emergence of the *Fourth Industrial Revolution*, connectivity is increasingly becoming a must-have for countries as they develop their digital strategy; and (4) Despite the growing awareness of the importance of the environment, it is difficult for the T&T sector to

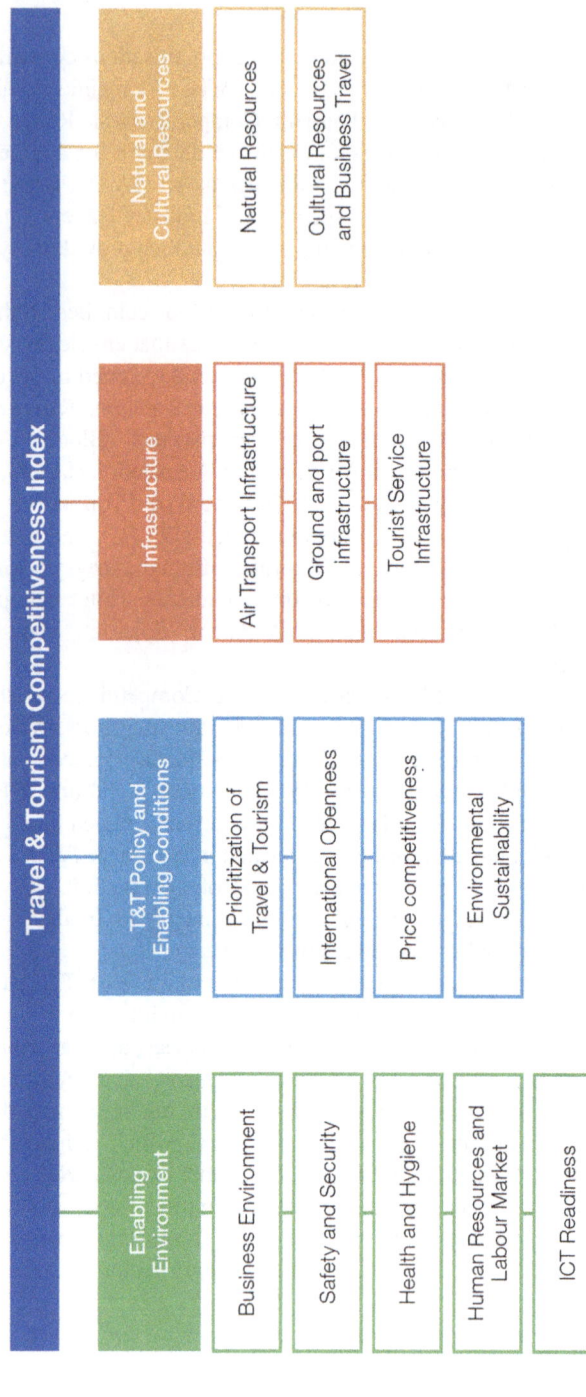

Fig. 2.1 The subindices and pillars of the T&TCI produced by the WEF (World Economic Forum 2017)

achieve sustainable development, given the degradation of natural resources it often entails.

Although it has been recognised as a useful tool for managing tourist destinations, the T&TCI elaborated by the WEF has also attracted a fair amount of criticism (Croes and Kubickova 2013). In particular, criticism has been levelled at the fact that differences across countries in terms of size and level of development are not accounted for in the construction of the index. Likewise, the criterion for selecting raw indicators has been questioned, and attention has been drawn to the problems that could arise as a result of combining quantitative and qualitative variables into a single index. In addition, critics have pointed out the arbitrariness in determining the weights used in the different aggregation processes involved in the calculation of the WEF T&TCI (Pulido-Fernández and Rodríguez-Díaz 2016). In this regard, the use of simple (unweighted) averages might not be appropriate when the raw indicators are not expected to have the same effect on competitiveness. Moreover, the pillars used by the WEF are made up of different numbers of indicators, ranging from 3 to 12. In practice, this means that some indicators contribute more to the T&TCI than others.

In our empirical application, we analyse 136 tourist destinations using 87 out of the 90 raw indicators provided by the WEF. The road density, paved road density and malaria indicators have been dropped for several reasons. For the first two indicators, information is not provided in the dataset due to proprietary rights. As far as the malaria indicator is concerned, it is measured as a categorical variable and so it is difficult to differentiate zeros from other values such as 'malaria free' or 'disappeared without specific measures'. Furthermore, some countries have no data for particular indicators. In these cases—which represent only 1.8% of the total—we have imputed the missing values by taking the average value for the same indicator from the countries ranked immediately above and immediately below the country in question in the relevant pillar (see technical details in Gómez-Vega and Picazo-Tadeo 2019).

That said, all raw indicators have been normalised to a scale of 1–7 according to the max-min or min-max formulas, depending on whether they are associated with greater or lesser competitiveness. Indicators for which higher scores imply higher T&TC, such as the number of cultural sites, have been normalised using the max-min formula as follows:

$$6\left(\frac{\text{Indicator}_{vd} - \text{Min}(\text{Indicator}_{vd})}{\text{Max}(\text{Indicator}_{vd}) - \text{Min}(\text{Indicator}_{vd})}\right) + 1 \qquad (2.1)$$

where Indicator_{vd} stands for the observed value of indicator v in tourist destination d.

On the other hand, for those indicators, such as terrorism incidence, which are negatively associated with T&TC, the min-max formula has been applied as follows:

$$6\left(\frac{\text{Indicator}_{vd} - \text{Max}(\text{Indicator}_{vd})}{\text{Min}(\text{Indicator}_{vd}) - \text{Max}(\text{Indicator}_{vd})}\right) + 1 \qquad (2.2)$$

Finally, these normalised indicators have been used to calculate the two T&TCI mentioned in the Introduction.

2.4 Methodology

Composite indicators are useful tools that synthesise in a single figure the information provided by multidimensional phenomena. While the OECD (OECD-JRC 2008) provides a general guideline for the construction of composite indicators, Mendola and Volo (2017) summarise the methodological foundations of composite indicators of T&TC, evaluating a set of currently available approaches and empirical applications. The paper provides scholars and practitioners with a set of statistical guidelines to build composite indicators and an operative scheme to assess indicators' effectiveness in empirical evaluations. In this regard, the authors propose a number of criteria to be observed when building synthetic indicators, which are based on previous literature (Munda and Nardo 2009; Paruolo et al. 2013; Nardo et al. 2005).

2.4.1 The DEA-MCDM Approach (Gómez-Vega and Picazo-Tadeo 2019)

Gómez-Vega and Picazo-Tadeo (2019) have recently proposed, as mentioned above, the computation of a T&TCI using DEA and MCDM. DEA techniques were originally proposed by Charnes et al. (1978) as a way to measure performance; they were later adapted for use in computing composite indicators by Lovell et al. (1995). This adaptation requires the assumption of the existence of an input—set equal to one—that is available to all tourist destinations; this input works as a *helmsman*. All the individual indicators of T&TC are considered as outputs. In our case, we observe $v = 1, \ldots, 87$ indicators representing different competitiveness dimensions (outputs) of $d = 1, \ldots, 136$ tourist destinations. The simplest formulation that allows the computation of a T&TCI for destination d' is:

$$\text{T\&TCI}_{d'}^{\text{DEA}} = \text{Maximise}_{w_{vd'}} \sum_{v=1}^{87} w_{vd'} \text{Indicator}_{vd'} \qquad (2.3)$$

Subject to:

$$\sum_{v=1}^{87} w_{vd'}\text{Indicator}_{vd} \leq 1 \quad d = 1, \ldots, 136$$

$$w_{vd'} \geq 0 \quad v = 1, \ldots, 87$$

Let us recall that Indicator$_{vd}$ stands for the observed value of indicator v in tourist destination d; moreover, w_{vd} is the weight assigned to that indicator. The T&TCI obtained from expression (2.3) are normalised to one, so that the higher the score, the higher the competitiveness. A noteworthy feature of DEA techniques is that the weights assigned to the individual indicators of competitiveness are endogenously computed at the tourist destination level. Moreover, in line with the abovementioned BoD principle, weights—which by construction are idiosyncratic—are set so as to maximise the competitiveness of each tourist destination when compared to all other destinations in the sample with the same weighting structure (Cherchye et al. 2007). Thus, indicators in which a tourist destination performs poorly will be assigned lower weights than those in which it displays better performance.

The empirical literature has shown DEA to be a successful approach for computing composite indicators involving a broad array of economic, environmental and social issues (see Zhou et al. 2007). However, it might not be so effective when it comes to establishing rankings. There are several reasons for this, among which two are particularly relevant in our case study: (1) a lack of discriminating power, which could prevent all tourist destinations in the sample from being fully ranked (see technical details in Dyson et al. 2001); and (2) the idiosyncratic nature of weights—which are different for each destination—might prevent meaningful comparisons of T&TC (technical explanations are provided in Kao and Hung 2005). In order to overcome both shortcomings, Gómez-Vega and Picazo-Tadeo (2019) combined DEA with MCDM, as proposed by Despotis (2002, 2005). In essence, this joint approach, which we will refer to as DEA-MCDM, consists of computing a common set of weights across tourist destinations for all the dimensions of T&TC, which allows direct comparisons and full rankings. This set of common weights is obtained from the linear program:

$$\text{Minimise}_{m_d, w_v, h} \frac{t}{136} \sum_{d=1}^{136} m_d + (1 - t)h \tag{2.4}$$

Subject to:

$$\sum_{v=1}^{87} w_v \text{Indicator}_{vd} + m_d = \text{T\&TCI}_d^{\text{DEA}} \quad d = 1, \ldots, 136$$

$$0 \leq m_d \leq h \quad d = 1, \ldots, 136$$

$$w_v \geq \varepsilon \quad v = 1, \ldots, 87$$

$$0 \leq h \text{ and } 0 \leq t \leq 1$$

In expression (2.4), w_v represents the common weight—across tourist destinations—assigned to indicator v; $T\&TCI_d^{DEA}$ is the DEA T&TCI obtained as the optimal solution of program (2.3) for destination d; ε is a non-Archimedian small number that guarantees that all individual indicators have a positive weight (0.001 in our case study); h is a non-negative parameter calculated by the program; m_d represents the difference between the T&TCI for destination d calculated with the DEA and DEA-MCDM approaches; and t is a pre-set parameter ranging between 0 and 1 that can be used to account for alternative theoretical assessments (see Bernini et al. 2013). When this parameter is equal to one—as assumed in our empirical application—the objective function to be minimised in expression (2.4) is the average difference across tourist destinations between DEA and DEA-MCDM T&TCI.

Finally, again following Gómez-Vega and Picazo-Tadeo (2019), the conventional DEA model in expression (2.3) is replaced by the additive DEA output-oriented Slacks Based Measure (SBM) proposed by Tone (2001). In this regard, Reig-Martínez et al. (2011) suggested the use of the more comprehensive SBM model of performance since it allows the researcher to account for both radial and non-radial (slacks) potential improvements in the dimensions of competitiveness, thus offering a more complete picture of performance. Formally, the SBM measure of competitiveness for tourist destination d' is given by the following program:

$$T\&TCI_{d'}^{SBM} = \text{Minimise}_{\lambda_d, s_{vd'}^+} \frac{1}{1 + \frac{1}{87}\left(\sum_{v=1}^{87} \frac{s_{vd'}^+}{\text{Indicator}_{vd'}}\right)} \tag{2.5}$$

Subject to:

$$\sum_{d=1}^{136} \lambda_d \leq 1$$

$$\text{Indicator}_{vd'} = \sum_{d=1}^{136} \lambda_d \text{Indicator}_{vd} - s_{vd}^+ \quad v = 1, \ldots, 87$$

$$0 \leq s_{vd}^+ \quad v = 1, \ldots, 87; \quad d = 1, \ldots, 136$$

$$0 \leq \lambda_d \quad d = 1, \ldots, 136$$

where s_{vd}^+ represents the slack obtained for indicator v and destination d, and λ_d measures the intensity with which tourist destination d enters the reference set to which tourist destination d' is compared.

2.4.2 The TOPSIS Model

TOPSIS is the acronym for 'Technique for Order Preferences by Similarity to Ideal Solutions', introduced by Hwang and Yoon (1981). This approach involves calculating two reference points: the positive ideal solution (PIS) and the negative ideal solution (NIS). The composite indicator is then obtained as the relative distance between each observation in the sample—the 136 tourist destinations in our empirical application—and these two reference points. Thus, observations are ranked according to how close they are to the PIS and how far they are from the NIS. The TOPSIS method is similar to the Grey Relational Analysis (GRA) proposed by Deng (1989), and since its initial formulation has been extended and adapted to different fields of study (Wang et al. 2017).

In their review of the literature on the use of MCDM techniques in the construction of composite indicators, El Gibari et al. (2019) find that TOPSIS is still the most widely adopted approach, accounting for 24% of cases. This quantification is based on the results of a bibliometric study of papers published after 2002 in international journals included in the Journal Citation Report. The authors present five categories of composite indicators, with TOPSIS included in the distance function approaches. They conclude that the distance function methods are still the most popular, noting an increasing trend since 2012.

Behzadian et al. (2012) analyse the *state-of-the-art* of TOPSIS by reviewing 266 related studies published in 103 journals since 2000. The authors find nine different fields of application, with 'supply chain management and logistics' and 'design, engineering, and manufacturing' accounting for around a half of the reviewed studies. Surprisingly, only a couple of the reviewed papers deal with tourism competitiveness: Huang and Peng (2012) and Zhang et al. (2011). In spite of this, TOPSIS continues to be very popular in T&TC analyses, as shown by 2 recent studies: Gu et al. (2019), which analyses tourism competitiveness in 13 cities located in the Sichuan province of China; and Sharma et al. (2019), focused on T&TC in the Indian States.

Turning to the formal issues of the TOPSIS approach to building composite indicators, the two reference points or ideal solutions mentioned above are computed as:

$$PIS = \{(\text{MaxIndicator}_{vd}/d = 1, \ldots, 136), v = 1, \ldots, 87\} \qquad (2.6)$$

$$NIS = \{(\text{MinIndicator}_{vd}/d = 1, \ldots, 136), v = 1, \ldots, 87\} \qquad (2.7)$$

The distances of a tourist destination d to these positive and negative ideal solutions are determined, respectively, as (Hwang and Yoon 1981; Zeleny 1998):

$$S_d^+ = \text{Dist}(\text{Indicator}_{vd}, \text{PIS}) = \sqrt{\sum_{v=1}^{87} (\text{Indicator}_{vd} - \text{PIS}_v)^2} \qquad (2.8)$$

$$S_d^- = \text{Dist}(\text{Indicator}_{vd}, \text{NIS}) = \sqrt{\sum_{v=1}^{87} (\text{Indicator}_{vd} - \text{NIS}_v)^2} \qquad (2.9)$$

Finally, the TOPSIS T&TCI for tourist destination d is calculated as:

$$\text{T\&TCI}_d^{\text{TOPSIS}} = \frac{S_d^-}{S_d^+ + S_d^-} \qquad d = 1, \dots, 136 \qquad (2.10)$$

This indicator ranges from 0 to 1, which higher scores indicating higher T&TC.

2.5 Results and Discussion

This section presents and discusses the results for the T&TCI of the 136 tourist destinations in our sample, computed with DEA-MCDM and TOPSIS. Table 2.1 displays the normalised weights assigned by the DEA-MCDM approach to the 14 pillars included in the analysis. The three most important pillars are cultural resources and business travel (0.257), air transport infrastructure (0.153) and natural resources (0.11); whereas the three least important ones are human resources and labour market (0.006), safety and security (0.016) and business environment (0.024).

The abovementioned figures confirm the results reported by Gómez-Vega and Picazo-Tadeo (2019) as, generally speaking, tourist destinations with outstanding

Table 2.1 Optimal weights for pillars in the DEA-MCDM T&TCI

Pillar	Weight
Business environment	0.024
Safety and security	0.016
Health and hygiene	0.026
Human resources and labour market	0.006
ICT readiness	0.043
Prioritisation of Travel & Tourism	0.080
International openness	0.065
Price competitiveness	0.052
Environmental sustainability	0.029
Air transport infrastructure	0.153
Ground and port infrastructure	0.061
Tourist service infrastructure	0.080
Natural resources	0.110
Cultural resources and business travel	0.257

Table 2.2 TOPSIS T&TCI ideal solutions

Pillar	PIS	Country	NIS	Country
Business environment	6.16	Hong Kong SAR	2.43	Venezuela
Safety and security	6.65	Finland	2.59	Colombia
Health and hygiene	6.86	Germany	1.83	Mozambique
Human resources and labour market	5.76	Iceland	2.56	Mauritania
ICT readiness	6.47	Hong Kong SAR	1.57	Burundi
Prioritisation of Travel & Tourism	6.18	Malta	1.89	Congo, Democratic Republic
International openness	5.21	Singapore	1.32	Yemen
Price competitiveness	6.66	Iran, Islamic Republic	2.81	Switzerland
Environmental sustainability	5.80	Switzerland	2.78	Yemen
Air transport infrastructure	6.76	Canada	1.30	Lesotho
Ground and port infrastructure	6.40	Hong Kong SAR	1.79	Congo, Democratic Republic
Tourist service infrastructure	6.67	Austria	1.84	Burundi
Natural resources	6.13	Brazil	1.60	Moldova
Cultural resources and business travel	6.94	China	1.02	Lesotho

endowments of natural and cultural resources enjoy a competitive advantage (Cracolici et al. 2008; Crouch 2011; Crouch and Ritchie 1999; Kayar and Kozak 2010; Gómez-Vega and Herrero-Prieto 2018). Regarding air transport infrastructure, Graham and Dobruszkes (2019) analyse the important role played by air transport in tourism competitiveness, not only for long-haul tourists, who might not have any alternative means of transport but also for short-haul tourists, who have benefited from the expansion of low-cost carriers. The authors contend that good air accessibility is a necessary condition for tourist competitiveness.

On the other hand, it is more difficult to justify the lower importance assigned to other pillars, especially those related with safety and the business environment (see Gómez-Vega and Herrero-Prieto 2018). However, it is true that tourism is usually associated with low-skilled labour; e.g., Santos and Varejão (2007; p. 227) argue that tourism is characterised by a high proportion of women in the workforce, as in other low-paying industries. The authors identify four characteristic features of employment in the Portuguese T&T industry: (1) female employment dominates; (2) relatively low levels of schooling prevail; (3) the distribution of employment is skewed towards the lower end of the scale of skills; and (4) average job tenure is short. Heath (2002) comments on the UNWTO's vision 2020 for tourism, highlighting the importance of safety and security as tourism will not be possible in destinations experiencing civil turmoil, at war, or where tourists' health or security is threatened.

Table 2.2 displays the ideal solutions obtained by applying the TOPSIS approach. In this regard, the PIS is mostly represented by developed countries. Hong Kong SAR represents the PIS for three pillars: business environment, ICT readiness and

Table 2.3 Top 10 countries according to the TOPSIS T&TCI: Comparison with DEA-MCDM and WEF

Country	TOPSIS	DEA-MCDM	WEF
Spain	1	3	1
Japan	2	6	4
Germany	3	5	3
France	4	4	2
United States	5	1	6
United Kingdom	6	7	5
Australia	7	9	7
Canada	8	12	9
Hong Kong SAR	9	16	11
Italy	10	8	8

ground and port infrastructure. Other notable findings are that Malta represents the PIS for the prioritisation of travel and tourism, and Iran for price competitiveness. On the other hand, the NIS is represented by African countries in most cases. The NIS for international openness and environmental sustainability is represented by Yemen, while airport infrastructure and cultural resources and business travel—two of the most important pillars of competitiveness, according to the DEA-MCDM approach—are represented by Lesotho. It is also worth highlighting the cases of Venezuela and Switzerland, which represent the NIS for business environment and price competitiveness, respectively.

On the one hand, regarding the countries that most commonly represent the PIS, Hong Kong SAR received 26.6 million international tourists in 2016 (World Economic Forum 2017), and its T&T industry contributed 8% of national GDP and 8.6% of total employment. Hong Kong SAR can be seen as an exemplar of the most developed part of Asia, displaying world-class infrastructure and the top-ranked performance in ICT. Conversely, its price competitiveness and poor endowment of natural and cultural resources bring down its position in the international T&TC ranking.

On the other hand, with respect to the countries that most often represent the NIS, Yemen and Lesotho received 366.7 and 320 thousand international tourist arrivals in 2016, respectively; the share of their T&T industries in GDP and employment, respectively, reached 4.4% and 2.9% in Yemen, and 5.8% and 5.1% in Lesotho. Like other sub-Saharan African countries, Lesotho, which is landlocked by South Africa, is betting on the development of tourism—predominantly based on nature, adventure and sightseeing—as a way to promote economic growth and better living conditions for its population (Noome and Fitchett 2019; Rogerson and Letsie 2013; Yiu et al. 2015). In the 1990s, its T&T industry was mainly related to casinos in Maseru and Bophuthatswana, a more accessible city for the key South African markets of Johannesburg and Pretoria (Rogerson and Letsie 2013).

That said, the top 10 tourist destinations according to the TOPSIS T&TCI are Spain, Japan, Germany, France, United States, United Kingdom, Australia, Canada, Hong Kong SAR and Italy, in this order (Table 2.3); i.e., developed countries, most of them European. According to the World Economic Forum (2017), Europe

Table 2.4 Worst 10 countries according to the TOPSIS T&TCI: comparison with DEA-MCDM and WEF

Country	TOPSIS	DEA-MCDM	WEF
Mozambique	127	117	122
Cameroon	128	129	126
Sierra Leone	129	132	131
Nigeria	130	127	129
Mali	131	124	130
Mauritania	132	131	132
Burundi	133	135	134
Yemen	134	136	136
Congo, Democratic Republic	135	133	133
Chad	136	134	135

continues to be the most competitive region worldwide thanks to its cultural richness, the excellence of the tourism service providers, its international openness as well as its perceived safety. This continent also remains the largest T&T market in the world, with more than 600 million international arrivals, almost double the share of Asia-Pacific in 2016. Furthermore, the abovementioned ranking of top tourist destinations according to the TOPSIS T&TCI is fairly similar to that obtained from the DEA-MCDM approach, as well as the ranking provided by the WEF. Of the top 10 countries as ranked by TOPSIS, only Canada and Hong Kong SAR are not in this leading group according to DEA-MCDM T&TCI, and just Hong Kong SAR is left out of by the WEF. However, these countries still occupy very high positions in these rankings.

The ten least competitive tourist destinations according to their TOPSIS T&TCI are, in this order, Mozambique, Cameroon, Sierra Leone, Nigeria, Mali, Mauritania, Burundi, Yemen and Congo (Table 2.4); that is, mainly sub-Saharan countries. Although the performance of the T&T industries of some countries in the region has improved, sub-Saharan Africa still remains the least competitive region in the world (World Economic Forum 2017). Although it offers promising possibilities and has a large endowment of natural and cultural resources, the area only attracted 29 million international tourists in 2015; air connectivity and travel costs are seen as the main barriers to developing its full potential. The tail of least competitive tourist destinations is fairly stable regardless of the method used to calculate T&TC. There are only 2 destinations in the TOPSIS worst 10 that are not labelled as such by either the DEA-MCDM T&TCI or the WEF: Mozambique and Mali, which rank 117th and 124th according to the DEA-MCDM approach; and Mozambique and Cameroon, which are placed 122th and 126th in the ranking of worst performers by the WEF.

Figures 2.2, 2.3, and 2.4 offer a graphical illustration of the pairwise correlations between the rankings of T&TC obtained with the TOPSIS and DEA-MCDM approaches, as well as that provided by the WEF. At first sight, all three figures show how the correlations at the head and the tail of the distributions are much stronger than in the middle. Beyond this visual assessment, the Spearman's rank

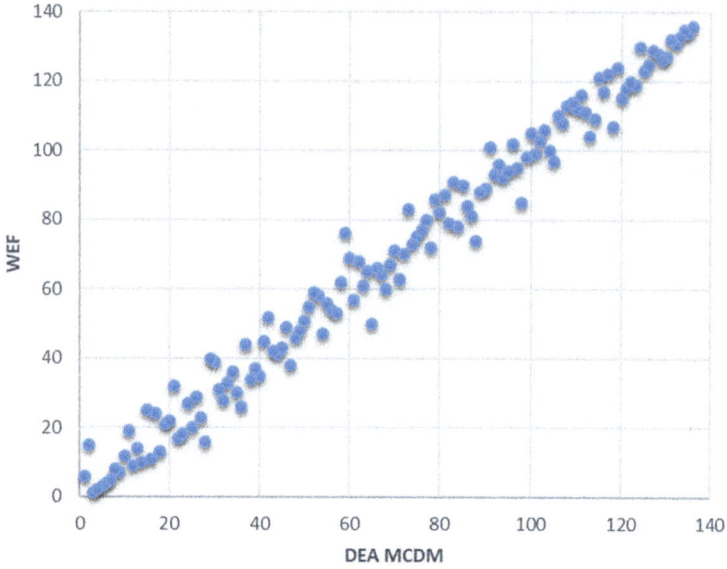

Fig. 2.2 Comparison between T&TCI: DEA-MCDM and WEF

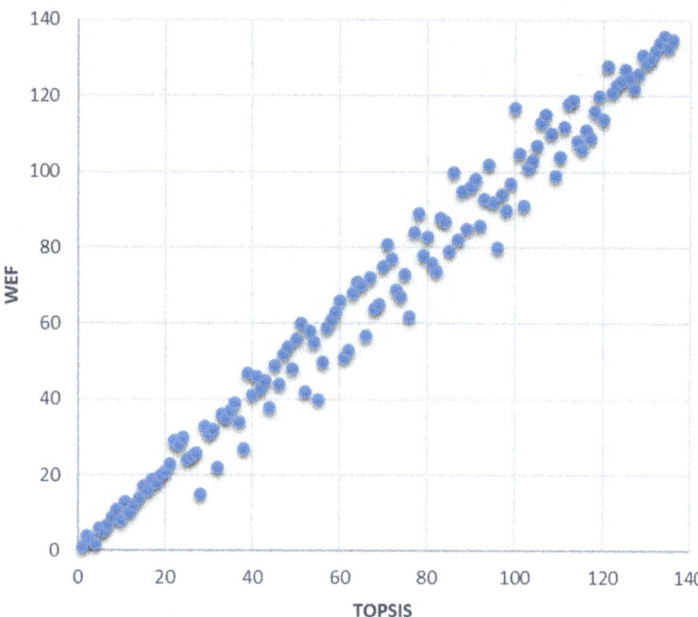

Fig. 2.3 Comparison between T&TCI: TOPSIS and WEF

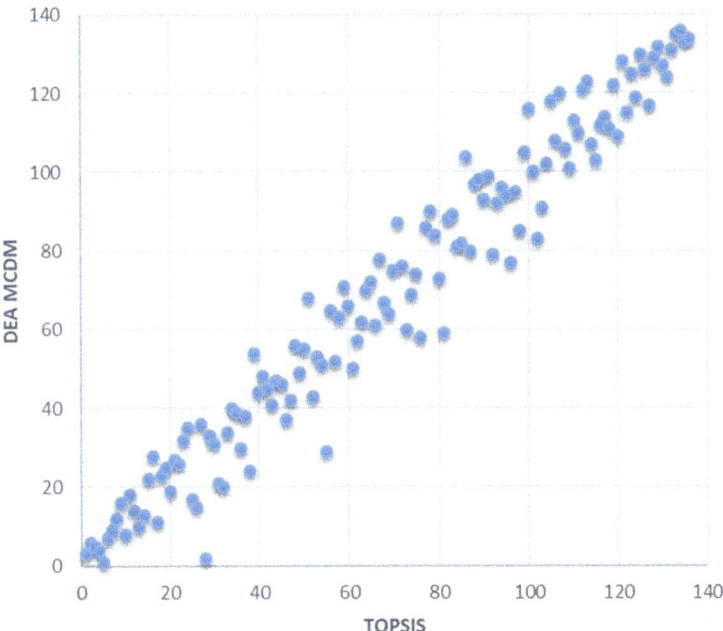

Fig. 2.4 Comparison between T&TCI: TOPSIS and DEA-MCDM

correlation coefficients unequivocally show that all pairwise correlations across rankings from the three approaches are positive and statistically significant at a 1% confidence level—coefficients between the rankings from T&TCI depicted in Figs. 2.2, 2.3, and 2.4 are 0.977, 0.989 and 0.990, respectively.

As the correlation tests mentioned above focus on the differences between rankings, it seems worth highlighting the countries that show the largest differences in position between the TOPSIS T&TCI and DEA-MCDM T&TCI. In this respect, the DEA-MCDM approach assigns the following countries a position at least 10 places higher than in the TOPSIS ranking: China and India (26 positions); Dominican Republic (22); Kenya and Tanzania (19); Colombia (18); Brazil (14); Jamaica, Guatemala and Honduras (13); Mexico, Cambodia and Uganda (12); and Iceland, Peru and Zimbabwe (11). Conversely, TOPSIS scores the following countries more generously than DEA-MCDM: Kuwait (18 positions); Bahrain (17); Kazakhstan and Moldova (16); Qatar (15); Tajikistan and Kyrgyz Republic (13); New Zealand, Saudi Arabia, and Macedonia (12); and Taiwan (China) and Montenegro (11).

The cases of India and China, the two tourist destinations that are most favoured by the DEA-MCDM approach, are interesting to analyse. According to WEF forecasts (World Economic Forum 2017), these two countries will lead the group of fastest growing tourist destinations in the period 2011–2031. They also lead the way in the expansion of the global middle class. The role of these two giants as a source of tourists is also remarkable, as only 5% of Chinese nationals currently have passports. India benefits substantially from air transport reforms designed to foster T&TC, which have enhanced connectivity within the country, and it is expected to become the third largest market in the world in the near future. It registered eight million international arrivals in 2015, whereas China received 57 million international tourists in 2016, a figure that accounts for over 20% of international arrivals in the region. It is interesting to observe that in the WEF ranking, China and India are positioned 15th and 40th, while according to the DEA-MCDM T&TCI they rank 2nd and 29th, respectively.

Finally, countries that fare significantly better under the TOPSIS approach are not particularly remarkable in terms of size, competitiveness or potential; among them, New Zealand and Taiwan—which received 3 and 10.4 million international arrivals in 2016, respectively—are the only tourist destinations that are relatively competitive according to the WEF, scoring 16th and 30th in the ranking, respectively. These countries are ranked 16th and 24th, respectively, according to the TOPSIS T&TCI.

2.6   Conclusions

This chapter compares the travel and tourism competitiveness (T&TC) of 136 tourist destinations, assessed using two widely accepted multi-criteria-decision-making (MCDM) methods: Data Envelopment Analysis (DEA) and the Technique for Order of Preference by Similarity to Ideal Solutions (TOPSIS). The 2017 edition of the Travel and Tourism Competitiveness Report published by the World Economic Forum (2017) is used as the main source of data. The two approaches applied rely on different conceptualisations. DEA-MCDM is based on obtaining a common set of weights to build a composite index of travel and tourism competitiveness (T&TCI), while TOPSIS relies on the analysis of how close a tourist destination is to the so-called positive ideal solution and how far it is from the negative ideal solution.

The main results obtained from the DEA-MCDM approach suggest that the three most important pillars of T&TC are, in this order, cultural resources and business travel, air transport infrastructure and natural resources. Whereas the first two are more discretionary in the sense that managers and policymakers can develop appropriate strategies to improve them, the latter constitutes a natural endowment. Regarding cultural resources and business travel, the top 5 tourist destinations worldwide are China, Spain, France, Japan and Italy; these countries enjoy a competitive advantage in several indicators: (1) number of world heritage cultural sites; (2) oral traditions and expressions as intangible cultural heritage; (3) number of large sports stadiums; (4) number of international association meetings; and (5) cultural and entertainment tourism digital demand. Richards (2001; p. 4) contends that 'cultural attractions play an important role in tourism at all levels, from the global highlights of world culture to attractions that underpin local identities. At a global level, cultural attractions are often seen as icons of important streams of global culture. This idea is enshrined in the designation of World Heritage Sites, or . . . . the designation of the Cultural Capital of Europe each year'. On the other hand,

Falk and Hagsten (2018) suggest that European cities' cultural attractions also play a key role in their ability to host international conferences.

Our results also show that the head and the tail of the distributions of the computed T&TCI seem to be more stable in terms of the destinations included, and thus less sensitive to the approach employed to assess competitiveness. Conversely, the main differences between the DEA-MCDM and TOPSIS approaches are observed in the middle of the distributions of T&TC. In this regard, there are winners and losers under the two approaches: China and India are paradigmatic examples of countries that score better under DEA-MCDM than TOPSIS, while New Zealand and Taiwan are the destinations that are most favoured by the TOPSIS approach.

To conclude, there are multiple MCDM models that can be employed to assess T&TC, all of which have their pros and cons. Besides, further approaches will be developed in the years to come, in the search for more comprehensive and accurate measures of competitiveness. The two methods compared in this chapter are more challenging to apply than the T&TCI provided by the World Economic Forum (WEF). Nonetheless, looking beyond the complexity and validity of the methods, it is important that the results are clearly communicated to the stakeholders involved in the promotion of tourist destinations, as such indices can be invaluable tools helping decision-makers to identify the weaknesses of tourist destinations and act accordingly.

References

Bahar O, Kozak M (2007) Advancing destination competitiveness research: comparison between tourists and service providers. J Travel Tour Mark 22:61–71

Behzadian M, Otaghsara SK, Yazdani M, Ignatius J (2012) A state-of the-art survey of TOPSIS applications. Expert Syst Appl 39:13051–13069

Bernini C, Guizzardi A, Angelini G (2013) DEA-like model and common weights approach for the construction of a subjective community well-being indicator. Soc Indic Res 114:405–424

Bornhorst T, Ritchie JR, Sheehan L (2010) Determinants of tourism success for DMOs & destinations: an empirical examination of stakeholders' perspectives. Tour Manag 31:572–589

Charnes A, Cooper WW, Rhodes E (1978) Measuring the efficiency of decision-making units. Eur J Oper Res 2:429–444

Chen CY, Sok P, Sok K (2008) Evaluating the competitiveness of the tourism industry in Cambodia: self-assessment from professionals. Asia Pac J Tour Res 13:41–66

Cherchye L, Moesen W, Rogge N, van Puyenbroek T (2007) An introduction to "benefit of the doubt" composite indicators. Soc Indic Res 82:111–145

Cracolici MF, Nijkamp P, Rietveld P (2008) Assessment of tourism competitiveness by analysing destination efficiency. Tour Econ 14:325–342

Croes R (2011) Measuring and explaining competitiveness in the context of Small Island destinations. J Travel Res 50:431–442

Croes R, Kubickova M (2013) From potential to ability to compete: towards a performance-based tourism competitiveness index. J Destin Mark Manag 2:146–154

Crouch GI (2011) Destination competitiveness: an analysis of determinant attributes. J Travel Res 50:27–45

Crouch GI, Ritchie JB (1999) Tourism, competitiveness, and societal prosperity. J Bus Res 44:137–152

Das J, Dirienzo C (2010) Tourism competitiveness and corruption: a cross-country analysis. Tour Econ 16:477–492

De Vita G, Kyaw KS (2016) Tourism development and growth. Ann Tour 60:23–26

Deng JL (1989) Introduction to Grey system theory. J Grey Syst 1:1–24

Despotis DK (2002) Improving the discriminating power of DEA: focus on globally efficient units. J Oper Res Soc 53:314–322

Despotis DK (2005) A reassessment of the human development index via data envelopment analysis. J Oper Res Soc 56:969–980

Dwyer L, Cvelbar LK, Edwards D, Mihalic T (2012) Fashioning a destination tourism future: the case of Slovenia. Tour Manag 33:305–316

Dwyer L, Cvelbar LK, Mihalic T, Koman M (2014) Integrated destination competitiveness model: testing its validity and data accessibility. Tour Anal 19:1–17

Dyson R, Allen R, Camanho A et al (2001) Pitfalls and protocols in DEA. Eur J Oper Res 132:245–259

El Gibari S, Gómez T, Ruiz F (2019) Building composite indicators using multicriteria methods: a review. J Bus Econ 89:1–24

Falk M, Hagsten E (2018) The art of attracting international conferences to European cities. Tour Econ 24:337–351

Gómez-Vega M, Herrero-Prieto LC (2018) Achieving tourist destination competitiveness: evidence from Latin-American and Caribbean countries. Int J Tour Res 20:782–795

Gómez-Vega M, Picazo-Tadeo AJ (2019) Ranking world tourist destinations with a composite indicator of competitiveness: to weigh or not to weigh? Tour Manag 72:281–291

Graham A, Dobruszkes F (2019) Air transport: a tourism perspective. Elsevier, Oxford

Gu T, Ren P, Jin M, Wang H (2019) Tourism destination competitiveness evaluation in Sichuan province using TOPSIS model based on information entropy weights. Discret Cont Dyn Syst 12:771

Heath E (2002) Towards a model to enhance Africa's sustainable tourism competitiveness. J Public Adm 37:327–353

Huang JH, Peng KH (2012) Fuzzy Rasch model in TOPSIS: a new approach for generating fuzzy numbers to assess the competitiveness of the tourism industries in Asian countries. Tour Manag 33:456–465

Hwang CL, Yoon K (1981) Multiple attribute decision making: methods and applications. Springer, New York

Kao C, Hung HT (2005) Data envelopment analysis with common weights: the compromise solution approach. J Oper Res Soc 56:1196–1203

Kayar ÇH, Kozak N (2010) Measuring destination competitiveness: an application of the travel and tourism competitiveness index (2007). J Hosp Mark Manag 19:203–216

Lan LW, Wu WW, Lee YT (2012) Exploring an objective weighting system for travel & tourism pillars. Procedia Soc Behav Sci 57:183–192

Lee YJ (2015) Creating memorable experiences in a reuse heritage site. Ann Tour Res 55:155–170

Lovell CAK, Pastor JT, Turner JA (1995) Measuring macroeconomic performance in the OECD: a comparison of European and non-European countries. Eur J Oper Res 87:507–518

Mazanec JA, Ring A (2011) Tourism destination competitiveness: second thoughts on the World Economic Forum reports. Tour Econ 17:725–751

Mendola D, Volo S (2017) Building composite indicators in tourism studies: measurements and applications in tourism destination competitiveness. Tour Manag 59:541–553

Munda G, Nardo M (2009) Non-compensatory/non-linear composite indicators for ranking countries: a defensible setting. Appl Econ 41:1513–1523

Nardo M, Saisana M, Saltelli A, Tarantola S (2005) Tools for composite indicators building. European Commission-OECD, Brussels

Noome K, Fitchett JM (2019) An assessment of the climatic suitability of Afriski Mountain Resort for outdoor tourism using the Tourism Climate Index (TCI). J Mt Sci 16:2453–2469

OECD-JRC (2008) Handbook on constructing composite indicators. Methodology and user guide. OECD Publishing, Paris

Oklevik O, Gössling S, Hall C et al (2019) Overtourism, optimisation, and destination performance indicators: a case study of activities in Fjord Norway. J Sustain Tour 27:1804–1824

Paruolo P, Saisana M, Saltelli A (2013) Ratings and rankings: Voodoo or science? J R Stat Soc Ser A 176:609–634

Pérez-Moreno S, Rodríguez B, Luque M (2016) Assessing global competitiveness under multi-criteria perspective. Econ Model 53:398–408

Pulido-Fernández JI, Rodríguez-Díaz B (2016) Reinterpreting the World Economic Forum's global tourism competitiveness index. Tour Manag Perspect 20:131–140

Reig-Martínez E, Gómez-Limón JA, Picazo-Tadeo AJ (2011) Ranking farms with a composite indicator of sustainability. Agric Econ 42:561–575

Richards G (2001) Cultural attractions and European tourism. CABI, Oxon

Ritchie JR, Crouch GI (2003) The competitive destination. A sustainable tourism perspective. CAB International, Wallingford

Rogerson CM, Letsie T (2013) Informal sector business tourism in the global south: evidence from Maseru, Lesotho. Urban Forum 24:485–502

Sainaghi R, Phillips P, Zavarrone E (2017) Performance measurement in tourism firms: a content analytical meta-approach. Tour Manag 59:36–56

Santos LD, Varejão J (2007) Employment, pay and discrimination in the tourism industry. Tour Econ 13:225–240

Sharma K, Dash M, Majumder MG, Sharma MG (2019) Hospitality competitiveness index for Indian States and Union Territories using Multi-Criteria TOPSIS Model. In: Rezaei S (ed) Quantitative tourism research in Asia: current status and future directions. Springer, Singapore, pp 59–73

Tone K (2001) A slacks-based measure of efficiency in data envelopment analysis. Eur J Oper Res 130:498–509

Wang Q, Dai HN, Wang H (2017) A smart MCDM framework to evaluate the impact of air pollution on city sustainability: a case study from China. Sustainability 9:911

World Economic Forum (2017) World economic forum, Geneva

World Tourism Organization (2019a) International tourism highlights, 2019 ed. UNWTO, Madrid

World Tourism Organization (2019b) Tourism enjoys continued growth generating USD 5 billion per day. UNWTO. Press release 19050, Madrid

Yiu L, Saner R, Lee MR (2015) Lesotho, a tourism destination: an analysis of Lesotho's current tourism products and potential for growth. In: Camillo A (ed) Handbook of research on global hospitality and tourism management. IGI Global, Pennsylvania, pp 312–331

Zeleny M (1998) Multiple criteria decision making: eight concepts of optimality. Hum Syst Manag 17(2):97–107

Zhang H, Gu CL, Gu LW, Zhang Y (2011) The evaluation of tourism destination competitiveness by TOPSIS & information entropy—a case in the Yangtze river delta of China. Tour Manag 32:443–451

Zhou P, Ang BW, Poh KL (2007) A mathematical programming approach to constructing composite indicators. Ecol Econ 62:229–291

Chapter 3
Smart Tourism Specialization to Outfox the Competition: *An Analytical Framework*

Jorge Ridderstaat

Abstract This study provides an analytical framework of smart tourism specialization, integrating smart tourism, tourism specialization, and nations' competitive advantage. The resulting model incorporates a process of tourism specialization, distinguishing between envisioned and realized tourism specialization. The model's dynamic nature originates from the applied smart tourism management approach and a feedback loop while considering the possibility that tourism development is not an isolated process. The study contributes to the literature by analytically linking smart tourism with tourism specialization and nations' competitiveness in a smart tourism specialization framework. Also, the study propels two new concepts in smart tourism literature, i.e., smart tourism specialization and smart tourism management. The analysis framework serves as a blueprint for policymakers in their quest to outfox the competition.

Keywords Tourism · Specialization · Demand · Smart · Management · Competition

3.1 Introduction

The "smart" concept was born from the scientific and technological revolution experienced since the end of the twentieth century (Dexeus 2019) and describes technological, economic, and social developments that are impacted by technology (Gretzel et al. 2015a). The term, which generally drifts toward user-friendliness rather than intelligence (Nam and Pardo 2011), has expanded to the tourism phenomenon, with buzzwords such as "smart destinations" and "smart tourism." While smart destinations emphasize the enrichment of the tourist experience through embedded technology (Buhalis and Amaranggana 2014), smart tourism aims to develop information and communication infrastructure and capabilities to improve

J. Ridderstaat (✉)
Rosen College of Hospitality Management, University of Central Florida, Orlando, FL, USA
e-mail: Jorge.Ridderstaat@ucf.edu

© The Author(s), under exclusive licence to Springer Nature Singapore Pte Ltd. 2021   37
S. Suzuki et al. (eds.), *Tourism and Regional Science*, New Frontiers in Regional
Science: Asian Perspectives 53, https://doi.org/10.1007/978-981-16-3623-3_3

management and governance to facilitate innovation of service and/or products, to improve the tourism experience, and ultimately, the competitiveness of tourism companies and destinations (Gretzel et al. 2015a).

Tourism is, globally, an important source of economic development. According to the World Travel and Tourism Council (WTTC), travel and tourism contributed to about 10.3% of the global gross domestic product and 10.4% of total global employment (330 million jobs) in 2019 (WTTC 2020). One important indicator of tourism development is tourism demand, which has to do with consuming goods and services at a destination (Cooper et al. 1993; Frechling 1996, 2001). Tourism demand is key for business profitability (Song, Witt, and Li 2009), and its effect can work further in the destinations' overall economy. The latter gives rise to another key term frequently used in the tourism literature, i.e., tourism specialization. While being a popular term, our understanding of the specialization concept remains poor due to the implicit consideration in many studies that the term is associated with certain demand- and supply-side factors, such as the direct contribution of tourism receipts to the gross domestic product (GDP), the direct and indirect contribution of travel and tourism in the GDP, the contribution of tourism receipts in total exports, the number of visitors as a percent of the total population in a destination, room index, tourism penetration index, and shops per person tourist density ratio (as summarized by Croes et al. 2018; Pérez-Dacal et al. 2014). From this standpoint, tourism specialization is considered only a retrospective concept, i.e., only measurable after the tourism activity. A recent study by Croes et al. (2020) has considered tourism specialization as a dynamic concept where, besides an after-event concept, there is also a prospective notion of the specialization concept, i.e., a decision on how a destination organizes its economy with tourism activities. According to the authors, both viewpoints of tourism specialization are connected through a feedback loop, implying a constant learning process. The feedback approach indicates destinations' capacity to innovate and upgrade their tourism specialization process, which ultimately defines their competitive success.

Against the previously described background, the present study posits a smart tourism specialization conceptual framework that integrates smart tourism, tourism specialization, and nations' competitiveness. Specifically, the scheme emphasizes smart tourism management's role in achieving a competitive edge and the established goals of specialization. The study contributes to advancing the smart tourism literature by integrating smart tourism with tourism, tourism specialization, and nations' competitiveness in a smart tourism specialization framework. Mapping the connection between the three concepts could help policymakers understand the different considerations on the path of achieving and sustaining a competitive edge over other destinations. The study also propels two new concepts in smart tourism literature, i.e., smart tourism specialization and smart tourism management. While the first concept emphasizes a dynamic process of achieving a predefined level of tourism influence, the second one focuses on a leadership process to achieve both the competitive advantage of a destination and a predefined specialization goal.

The remainder of the study is as follows. The next section discusses the smart concept and its association with tourism. The smart tourism concept cannot be detangled from conventional elements of the tourism phenomenon, such as tourism demand and tourism specialization, which are also discussed in the section. The second segment also discusses the relevance of comparative advantage in explaining the smart tourism specialization objective. The study presents a schematic overview of the smart tourism specialization process in the third section, while the last segment concludes and considers managerial implications, limitations, and ideas for future research.

3.2 Literature Review

3.2.1 General

The world has experienced a new scientific and technological revolution since the end of the twentieth century, covering disciplines such as quantum physics, biomolecular chemistry, energy, materials, and information and communication technologies (Dexeus 2019). The "smart" concept is a buzzword to describe technological, economic, and social developments that are impacted by technology (e.g., sensors, big data, open data, and new connectivities and exchange of information) (Gretzel et al. 2015b). The term gravitates in practice more towards user-friendliness than intelligence (Nam and Pardo 2011). The smart notion has subsequently been adjoined with terminologies such as cities (smart cities), tourism destinations (smart tourism destinations), and tourism (smart tourism) to create other amalgams in the smart sphere.

3.2.2 Smart Cities, Smart Destinations, and Smart Tourism

Smart cities are an important output of the technological innovation revolution, where the goal is to make traditional networks and services more flexible, efficient, and sustainable for the benefit of residents (Mohanty et al. 2016). The goal is to greatly improve the living and working environment of those living in the city (Hall et al. 2000). There are many definitions of the smart city concept, and the meaning of a smart city is multifaceted, complicating our understanding of the concept (Albino et al. 2015). Nam and Pardo (2011), for example, distinguish between technological, human, and institutional dimensions, while Giffinger et al. (2007) consider the smart economy, smart people, smart governance, smart mobility, smart environment, and smart living as key domains of smart cities. Joss et al. (2019), based on a review of the literature and text string analyses, identified ten different smart city dimensions (sustainability, environment, experiment/innovation, spatial planning/development, economy, society, digital technology, international, infrastructure, and governance). The oscillating dimensional nature of smart cities explains, to some extent, the different meanings of this concept in the literature, as summarized by Albino et al. (2015).

The evolution of smart cities has also made it possible to create smart tourism destinations, which can be defined as "a city that embraces a culture of sustainability and excels at smart urban technologies and intelligent mobility solutions in order to present it as a 'smart tourism city'—a city that offers both visitors and residents an exciting yet relaxed, authentic, comfortable and 'green' urban experience." (Europaforum Wien –Center for Urban Dialogue and European Policy 2014: 6). However, while a city can be a tourist destination, a tourist destination is not always a city but can also be an area or a territory (Beritelli et al. 2007) or a place (UNWTO 2008). With technology embedded in all organizations and entities, smart tourism destinations can exploit the ensuing synergy effects to support the enrichment of tourists' experiences (Buhalis and Amaranggana 2014).

While smart tourism destinations aim to enrich the tourism experience, smart tourism has a broader connotation than just the destination's tourism impression. It generally aims to develop information and communication infrastructure and capabilities to improve management and governance to facilitate innovation of service and/or products, enhance the tourism experience, and, finally, improve the competitiveness of tourism companies and destinations (Gretzel et al. 2015a). Gretzel et al. (2015b: 181) define smart tourism as "tourism supported by integrated efforts at a destination to collect and aggregate/harness data derived from physical infrastructure, social connections, government/organizational sources and human bodies/minds in combination with the use of advanced technologies to transform that data into on-site experiences and business value-propositions with a clear focus on efficiency, sustainability and experience enrichment." While the concept of smart tourism is clear and seems to profess an emphasis on technological and data approaches aimed at accommodating the travelers' experience with the destination, it provides only an implicit approach to tourism demand. This latter is an important constituent of the tourism phenomenon and, in its turn, cannot be detached from the tourism specialization concept.

3.2.3   Tourism Demand and Tourism Specialization

Tourism demand is a key indicator of tourism activity and is associated with the classical explanation of people's desire to consume goods and services (Song et al. 2010). Several authors have contributed to defining tourism demand. For example, Cooper et al. (1993: 15) defined tourism demand as "the amount of any product or service which people are willing and able to buy at each specific price in a set of possible process during a specific period of time." Pearce (1995: 18) defined tourism demand as "the relationship between individuals' motivation [to travel] and their ability to do so." Frechling (1996: 3; 2001: 4) defined tourism demand as "a measure of visitors' use of goods and services." These definitions suggest that tourism demand is about goods and services that people wish and can buy, but are only implicit in the context in which these consumption take place. Of course, the consumption has to occur in a tourism setting, where the consumer is a tourist. The latter implies that the consumer has to travel to a destination outside their usual environment for the same day or longer, but still less than a year, with business, leisure, or other intentions, according to the United Nations' World Tourism Organization (UNWTO 2008).

Consumption of tourism needs to occur on-site (Croes 2006), which makes this activity important for destinations' development. Many businesses, such as tour operators, hotels, recreation facilities, and shop owners at the destination depend on tourism demand for their profitability (Song, Witt, & Li 2009). Besides business profitability, the effect of tourism demand can work further in the destinations' overall economy (macroeconomic impact). The latter depends on how specialized a country is in tourism (tourism specialization). While the concept of tourism specialization has been amply used in tourism studies, this notion's meaning has remained unambiguously elusive. The relevant literature has implicitly created the image that tourism specialization is already a generally understood concept that does not require further elaboration. Multiple studies have directly proceeded to define demand and supply-side indicators for conceptualizing tourism specialization, such as the direct contribution of tourism receipts in the gross domestic product (GDP), the direct and indirect contribution of travel and tourism in the GDP, the contribution of tourism receipts in total exports, the number of visitors as a percent of the total population in a destination, room index, tourism penetration index, and shops per person tourist density ratio (e.g., Croes et al. 2018; Pérez-Dacal et al. 2014). From this perspective, tourism specialization is seen mostly as an ex post concept, i.e., after tourism has occurred. However, as pointed out by Croes et al. (2020), tourism specialization can also be seen as a dynamic concept indicating how a country organizes its economy. A country can decide in advance that it wants to reorganize its economy in such a way to emphasize tourism in "a dynamic process integrating resources and assets, providing a sense of place and identity to tourism markets." (Croes et al. 2020: 3). According to the authors, the latter suggests a constant learning process (feedback) that requires the organization and coordination of several actors in creating and managing a unique experience where individuals' feelings and perceptions matter. Tourism specialization is likely to be a means to an end, whether the latter has intended or unintended consequences in, for example, the economic, social, environmental, and cultural spheres.

3.2.4 Theoretical Background: Specialization, Comparative and Competitive Advantage

Adam Smith (1723–1790), in his famous book "An Inquiry into the Nature and Causes of the Wealth of Nations" (Smith 1776; 2005), purported the specialization notion by introducing labor division as a means to increase the productive power of labor. Smith championed the notion of extending labor division beyond the domestic borders, i.e., an international division of labor (Cho and Moon 2013). In his "The Principles of Political Economy and Taxation" (Ricardo 1817; 2004), David Ricardo looked beyond the value of labor as a means of production. Rather, the earth's produce was the outcome of a united application of labor, machinery, and capital. Still, Ricardo considered labor as the foundation of the value of commodities. In his book mentioned previously, Ricardo propelled the notion of comparative advantage of nations using the example of cloth and wine trading between England and Portugal. While England requires 100 men to produce cloth and 120 men for wine, Portugal's labor input for both commodities is much less (90 for cloth and 80 for wine). One could say then that Portugal has an absolute advantage in producing both commodities because of its lower use of labor. However, producing both wine and cloth would require more labor in Portugal and would require diverting a portion of capital from the cultivation of vines to cloth manufacturing. Portugal can produce wine at 2/3 (80/120) of labor compared to England, and cloth at 90% (90/100) of labor vis-à-vis England. Comparatively, Portugal would be advantageous to export wine in exchange for cloth, notwithstanding that Portugal could produce a cloth with less labor than England. However, by concentrating on wine production, Portugal could obtain more cloth from England than local production. So, rather than looking at the absolute advantage of nations, Ricardo emphasized countries' comparative advantage. Although not explicitly mentioned, Ricardo's notion of comparative advantage was for each country to specialize in those commodities in which it had a favorable position vis-à-vis another country. Ricardo championed commodity specialization rather than labor specialization, as was the case of Adam Smith. However, Ricardo's theory seems to promote an extreme degree of specialization, i.e., where countries produce only one product, contrary to the multiproduct practice (Cho and Moon 2013).

Moreover, the comparative advantage of nations is likely impossible in all sectors of the economy because of an economy-wide resource constraint. As put by Krugman (1987: 42): "Japan cannot have a competitive advantage over the U.S. in everything, because if it did, there would be an excess demand for Japanese labor. Japanese relative wages would rise (perhaps via an exchange rate adjustment), which would restore U.S. competitiveness in some sectors." Ricardo was not the only one who put a theory forward on the comparative advantage of nations. In 1919 and his student Bertil Ohlin in 1933, Eli Heckscher created a theory on international trade, the so-called Heckscher-Ohlin theorem. According to Heckscher, countries will export commodities, which require relatively intensive use of the abundantly available production factors (Magnusson 1994). Ohlin provided the conditions for interregional exchange between countries, i.e., differences in factor endowments that lead to differences in comparative costs (Flam and Flanders 2000).

Contrary to the Ricardian understanding of comparative advantage, the Heckscher-Ohlin approach ignores differences in total factor productivity across industries and assumes that all countries possess the same production function in a given industry. That comparative advantage variation is derived from differences in factor abundance and the factor intensity of goods (Morrow 2010). Relatively similar countries, with few comparative differences, will experience no trade, but if there are economies of scale, there would be benefit from specialization by each country, according to the Heckscher-Ohlin theory (Cho and Moon 2013).

However, comparative advantage based on production factors is not sufficient to explain trade patterns, according to Porter (1990a, b). The author criticized the standard comparative advantage because it does not consider differences between nations stemming from variations in economies of scale, available technologies, manufactured products, and national factors. Also, the comparative advantage theorem fails to recognize that skilled labor and capital can move between countries. In Porter's own words, "At best, factor comparative advantage theory is coming to be seen as useful primarily for explaining broad tendencies in the pattern of trade (for example, its average labor or capital intensity) rather than whether a national exports or imports in individual industries." (Porter 1990a, b: 12). The notion of comparative advantage of nations may be too narrow and impossible to maintain, given that the globalization forces will allow companies to match comparative advantages of others by sourcing outputs of raw materials, capital, and generic scientific knowledge from anywhere around the world to take advantage of low-cost capital and labor, according to the author (Porter 2008). Consequently, it is a competitive advantage that is important, rather than comparative advantage.

Competitive advantage is a dynamic and evolving process, where the capacity of a nation's industry to innovate and upgrade plays a key role (Porter 1990b). Factors of production (labor, land, natural resources, capital, and infrastructure) are sustained, with heavy investment and specialized nature. To quote Porter (2008: 188) himself: "Nations succeed in industries where they are particularly good at factor creation. Competitive advantage results from the presence of world-class institutions that first create specialized factors and then continually work to upgrade them." The process of innovation and improvement depends on information, specifically those that are not available to competitors or that they do not pursue, according to the author (Porter 1990b). However, having the information is not enough. It matters how the information is being used to achieve a competitive advantage. Leadership plays an important role here. In the words of Porter (2008: 211): "Today's competitive realities demand leadership. Leaders believe in change, they energize their organizations to innovate continuously; they recognize the importance of their home country as integral to their competitive success and work to upgrade it. Most important, leaders recognize the need for pressure and challenge. Because they are willing to encourage appropriate—and painful—government policies and regulations, they often earn the title 'statesmen,' although few see themselves that way. They are prepared to sacrifice the easy life for difficulty and, ultimately, sustained competitive advantage."

The previous suggests that competitive advantage is more important than comparative advantage. It thrives in an environment where the factors of production are specialized in nature. Leaders have the appropriate (often exclusive) information and can energize others to innovate continuously. The dynamic and evolving nature of the competitive advantage scene requires a lively and adaptive consideration of the specialization-information-leadership connection, as proposed and discussed in the next section.

3.3   Smart Tourism Specialization: A Conceptual Bid

In the previous section, smart tourism was linked to technological and data approaches to add value to the travelers' experience. Furthermore, tourism specialization was seen as a dynamic process of resource concentration or economic reorganization. From a theoretical perspective, nations' competitive advantage was found to be a dynamic and evolving process, where countries' capacity to innovate and upgrade mattered for competitive success. We have learned that appropriate leadership is key to sustain continued innovations and upgrading, aided by (exclusive) information. Integrating smart tourism, tourism specialization, and competitive advantage of nations results in a conceptual framework of smart tourism specialization, as depicted in Fig. 3.1. In the gray area, smart tourism specialization is considered a dynamic process, as Croes et al. (2020) suggested. The process is continuous, starting from an envisioned level of involvement of a destination with tourism development. The goal here is to identify a vision of where the destination wants to be with its tourism development in the long-term future (Brent Ritchie 1993; OECD 2010). Once the vision has been determined, the next step is to develop the mission, which explains the pathway(s) to accomplish the vision (Edgell and Swanson 2019). The development of the vision and mission occurs in the ex ante stage of tourism specialization. The next step involves applying strategies, i.e., a different set of activities, as suggested by Porter (2008), to create a unique and valuable position. This stage is where smart tourism management becomes relevant. Smart in this context goes beyond the use of technology to impact technological, economic development, or the tourism experience and includes an intelligent and acumen approach, which assists in making good judgmental and quick decisions to outfox the competition. Collecting handy and perhaps exclusive information is not enough to achieve a competitive edge. Leadership is also important to use the available information to stimulate continuous innovation to stay ahead of the competition. Smart tourism management can allow destinations to achieve a competitive edge, which can ultimately benefit its tourism development, and, ultimately, the ex post stage of tourism specialization. In this phase, the outcome of tourism specialization is known and is most often implied by many studies, judging by the different indicators used in these studies.

With the available ex post information on tourism specialization, countries and destinations can now compare how close (or far) they are from achieving the envisioned tourism specialization goal through continuous feedback loops. The latter is important because it allows policymakers to assess whether the destination is on track to realize its envisioned tourism specialization objective and provide information on Smart Tourism Management's effectiveness. The feedback process in tourism specialization can have both a long-run and a short-run perspective. In

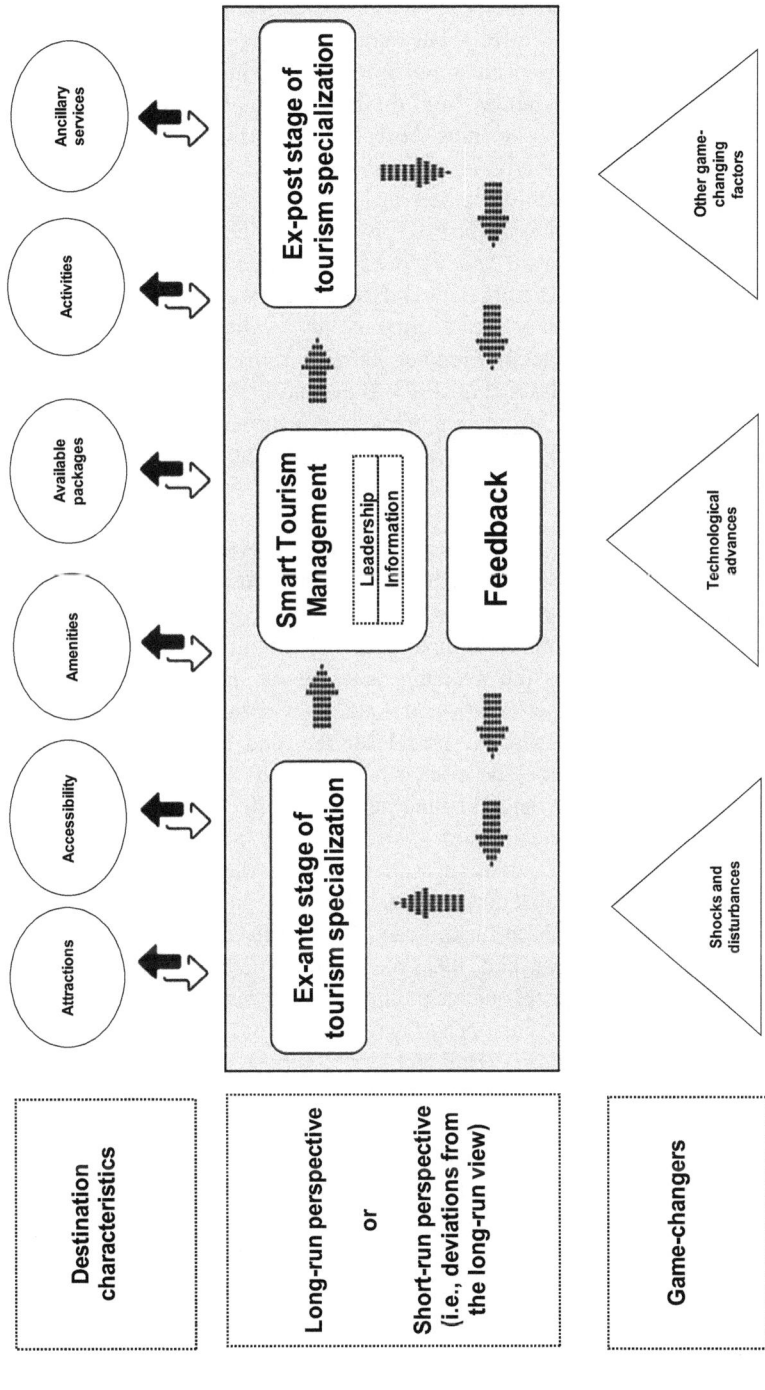

Fig. 3.1 Smart tourism specialization to outfox the competition

line with Ridderstaat et al. (2016a, b), the difference between the long- and short-run perspective should not be interpreted from a chronological perspective (i.e., first the short- and then the long-run effect), but rather in a simultaneous way. The latter implies that both perspectives occur simultaneously and that the short-run approach should be interpreted as deviations from the long-run movement. Data (including tourism-related information) can, thus, be collected at different timeframes (e.g., daily, weekly, monthly, quarterly, or annually), and, depending on the intensity of the data frequency, the information can include trend, seasonal, cyclical, and/or irregular patterns (Bails and Peppers 1993; Berenson et al. 2013). While the trend represents the long-run upward and downward movement in the data, the other elements represent deviations from the trend. Seasonal patterns represent deviations that repeat themselves every year, contrary to cyclical movements with a nonperiodic recurring character; irregular patterns indicate erratic and irregular changes in the data (Makridakis et al. 1983; Bails and Peppers 1993).

The smart tourism specialization process is not an endogenous (stand-alone) event and may be affected by (semi-)exogenous factors. One group of these factors has to do with the characteristics of the destination, and have been categorized by Buhalis (2000) and Buhalis and Amaranggana (2014) as the six A's of successful destinations, i.e., Attractions (natural, artificial, or human-made), Accessibility (the entire transportation system to reach and move around the destination), Amenities (all services facilitating a convenient stay), Available packages (availability of service bundles by intermediaries to direct tourists' attention to the unique features of a destination), Activities (all available activities at the destination and what consumers can do during their visit), and Ancillary Services (daily used services, such as banks, telecommunication, postal service, and hospital, which are not primarily aimed for tourists). These six A's require, to some extent, input from smart tourism management and, in their turn, can affect tourism demand and, ultimately, the tourism specialization goals. They are not fully controllable by smart tourism management, and, as such, they are semi-exogenous in nature.

The smart tourism specialization process can also be affected by what is called here, the game-changers. These factors can, temporarily or permanently, cause a shift in tourism development, and, thus, the tourism specialization process. They include shocks and disturbances and technological advances and refer to factors such as residents' quality of life. As an open system, tourism interacts with elements both inside and outside its boundaries (Hall and Lew 2009; McDonald 2009). The latter makes them susceptible to many shocks, including outbreaks of contagious diseases, terrorism, economic fluctuations, unstable currencies, energy crises, and climate change (Bonham et al. 2006; Butler 2009; Strickland-Munro et al. 2009). The COVID-19 pandemic, for example, has caused most countries around the world to close their borders for international travel to contain exposures to the virus but simultaneously caused vast declines in international tourism, estimated in the range of 60–80% for 2020 (OECD 2020). Technological advances and innovations can also affect the tourism specialization process. The smart concept discussed previously in this study has provided several examples of new data collection approaches (e.g., big data and open data) that may impact the way destinations market tourism. In the category of other game-changing factors, we can consider, for example, residents' quality of life as a determining element. Studies in recent years have shown that residents' pursuit of a better quality of life can also impact tourism destinations' success (e.g., Croes 2012; Ridderstaat et al. 2016a, b; Fu et al. 2020). The game-changing factors could, generally, have either a positive or a negative (and sometimes lasting) effect on a destination's tourism development and may be steered, to some extent, by smart destination management.

This section has proposed an integrated and dynamic approach to tourism specialization, i.e., by integrating the latter with the smart concept. The resulting smart tourism management concept is anchored on the technological and data principles conventionally found in smart tourism discussions. It integrates intelligence and acumen features of leadership to make swift and judgmental decisions to achieve the envisioned goals of tourism specialization. Smart tourism management is only part of the story, as the outcome of tourism development and, thus, tourism specialization can also be affected by the destination characteristics and the occurrence of game-changers. However, the smart tourism management approach could increase a destination's chances of achieving a competitive edge, enabling it to outfox the competition.

3.4 Concluding Remarks

3.4.1 General

This study proposed a conceptual scheme of smart tourism specialization that integrates smart tourism, tourism specialization, and destinations' competitive advantage. The resulting smart tourism specialization model incorporates a dynamic tourism specialization process, guided to some extent by a smart tourism management approach, where there is a feedback loop between envisioned and realized tourism specialization. Tourism development and tourism specialization do not occur in a sterile environment, and they may be affected by the characteristics of destinations and possible game-changers, such as shocks and technological advances.

Two new concepts associated with smart tourism can be distilled from the study's approach, i.e., smart tourism specialization and smart tourism management. Smart tourism specialization refers to a dynamic process of achieving a predefined level of tourism influence on, for example, the economic, social, cultural, and environmental sphere of destinations, using smart tourism management and prospective-retrospective information comparison, aimed at achieving a competitive edge over other destinations. Smart tourism management is considered a leadership process of planning, making decisions, and influencing other people, helped by technological innovations and information, to achieve a competitive edge and predefined tourism specialization objectives.

3.4.2 Managerial Implications

From a managerial stance, smart tourism specialization provides a blueprint of an integrated understanding of tourism development and tourism specialization, considering the open system nature of tourism and the factors to consider in achieving a competitive edge over other destinations (outfoxing the competition). As shown by the study, smart tourism management is a key input to outfox the competition, and appropriate information and leadership play important roles in achieving the latter. Policymakers can use the blueprint in their quest to achieve supremacy in tourism development over other competing destinations.

3.4.3 Limitations

Like many other studies, this one is not without limitations. The study presents a conceptual model as the first stage of an investigation. The empirical testing will occur in a later stage of the research. Furthermore, the study considered the model elements mostly as is where is and did not dive further into the smart tourism, tourism specialization, and competitiveness of nations, to further describe their integration mechanics.

3.4.4 Future Research

Future research should empirically test the presented conceptual scheme for validity. The idea is to use several countries as case studies to understand their approach towards tourism development and tourism specialization. Upcoming studies should also expand on the specialization, smart, and competitiveness concepts to further understand their integration into the presented conceptual framework. Future studies should also emphasize smart tourism management, specifically on their role in achieving the tourism specialization goals and their abilities to outfox the competition.

References

Albino V, Berardi U, Dangelico R (2015) Smart cities: definitions, dimensions, performance, and initiatives. J Urban Technol 22(1):3–21
Bails D, Peppers L (1993) Business fluctuations: forecasting techniques and applications. Prentice-Hall, Hoboken
Berenson M, Levine D, Szabat K (2013) Basic business statistics concepts and applications, 13th edn. Pearson, Boston

Beritelli P, Bieger T, Laesser C (2007) Destination governance: using corporate governance theories as a foundation for effective destination management. J Travel Res 46:96–107. https://doi.org/10.1177/0047287507302385

Bonham C, Edmonds C, Mak J (2006) The impact of 9/11 and other terrible global events on tourism in the United States and Hawaii. J Travel Res 45:99–110

Brent Ritchie J (1993) Crafting a destination vision. Tour Manag 14(5):379–389

Buhalis D (2000) Marketing the competitive destination of the future. Tour Manag 21:97–116

Buhalis D, Amaranggana A (2014) Smart tourism destinations. In: Ziang Z, Tussyadiah I (eds) Information and communication technologies in tourism. Springer International Publishing, Basel, pp 553–564

Butler R (2009) Tourism in the future: cycles, waves or wheels? Futures 41:346–352

Cho D, Moon H (2013) From Adam Smith to Michel Porter: evolution of competitiveness theory. World Scientific, Singapore

Cooper C, Fletcher J, Gilbert D, Wanhill S (1993) Tourism: principles and practice. Pitman, London

Croes R (2006) A paradigm shift to a new strategy for small island economies: embracing demand side economics for value enhancement and long term economic stability. Tour Manag 27 (3):453–465

Croes R (2012) Assessing tourism development from Sen's capability approach. J Travel Res 51 (5):542–554

Croes R, Ridderstaat J, Niekerk M (2018) Connecting quality of life, tourism specialization, and economic growth in small island destinations: the case of Malta. Tour Manag 65:212–223. https://academic.microsoft.com/paper/2765431042

Croes R, Ridderstaat J, Shapoval V (2020) Extending tourism competitiveness to human development. Ann Tour Res 80:102825

Dexeus C (2019) The deepening effects of the digital revolution. In: The future of tourism. Springer, pp 43–69

Edgell D, Swanson J (2019) Tourism policy and planning: yesterday, today, and tomorrow. Routledge, New York

Europaforum Wien – Centre for Urban Dialogue and European Policy (2014) Vienna tourism strategy 2020. Vienna Tourist Board, Vienna

Flam H, Flanders M (2000) The young Ohlin on the theory of "Interregional and international trade". Seminar paper no. 684. Institute for International Economic Studies, Stockholm University

Frechling D (1996) Practical tourism forecasting. Butterworth-Heinemann, Oxford

Frechling D (2001) Forecasting tourism demand: methods and strategies. Butterworth-Heinemann, Oxford

Fu X, Ridderstaat J, Jia H (2020) Are all tourism markets equal? Linkages between market-based tourism demand, quality of life, and economic development in Hong Kong. Tour Manag 77:1–13

Giffinger R, Fertner C, Kramar H, Meijers E (2007) City-ranking of European medium-sized cities. Cent Reg Sci Vienna UT:1–12

Gretzel U, Koo C, Sigala M, Xiang Z (2015a) Special issue on smart tourism: convergence of information technologies, experiences, and technologies. Electron Mark 25:175–177. https://doi.org/10.1007/s12525-015-0194-x

Gretzel U, Sigala M, Xiang Z, Koo C (2015b) Smart tourism: foundations and developments. Electron Mark 25:179–188

Hall C, Lew A (2009) Understanding and managing tourism impacts: an integrated approach. Routledge, New York

Hall R, Bouwerman B, Braverman J, Taylor J, Todosow H, Von Wimmersper U (2000) The vision of a smart city. In: 2nd international life extension technology workshop, Paris, France, September 28, 2000

Joss S, Sengers F, Schraven D, Caprotti F, Dayot Y (2019) The smart city as global discourse: Storylines and critical junctures across 27 cities. J Urban Technol 26(1):3–34

Krugman P (1987) The narrow moving band, the Dutch disease, and the competitive consequences of Mrs. Thatcher: notes on trade in the presence of dynamic scale economies. J Dev Econ 27 (1–2):41–55

Magnusson L (1994) Heckscher and mercantilism: an introduction. Uppsala papers in economic history research report no. 35

Makridakis S, Wheelwright S, McGee V (1983) Forecasting: methods and applications. Wiley, Hoboken

McDonald J (2009) Complexity science: an alternative view for understanding sustainable tourism development. J Sustain Tour 17(4):455–471

Mohanty S, Choppali U, Kougianos E (2016) Everything you wanted to know about smart cities: the internet of things is the backbone. IEEE Consumer Electron Mag 5(3):60–70

Morrow P (2010) Ricardian-Heckscher-Ohlin comparative advantage: theory and evidence. J Int Econ 82:137–151

Nam T, Pardo T (2011) Conceptualizing smart city with dimensions of technology, people, and institutions. In: The proceedings of the 12th annual international conference on digital government research

OECD (2010) Tourism trends and policies 2010. OECD, Paris

OECD (2020) Tourism policy responses to the coronavirus (COVID-19). Organization for Economic Cooperation and Development, Paris

Pearce DG (1995) Tourism today: a geographical analysis, 2nd edn. Longman Scientific & Technical, Harlow

Pérez-Dacal D, Pena-Boquete Y, Fernández M (2014) A measuring tourism specialization: a composite indicator for the Spanish regions. Almatourism J Tour Cult Territorial Dev 5 (9):35–73

Porter M (1990a) The competitive advantage of nations. The Free Press, New York

Porter ME (1990b) The competitive advantage of nations. Harv Bus Rev 68(2):73–93. (March–April 1990)

Porter ME (2008) On competition. Harvard Business Press, Boston, MA

Ricardo D (1817; 2004) The principles of political economy and taxation. Dover Publications, Inc., New York

Ridderstaat J, Croes R, Nijkamp P (2016a) A two-way causal chain between tourism development and quality of life in a small island destination: an empirical analysis. J Sustain Tour 24 (10):1461–1479

Ridderstaat J, Croes R, Nijkamp P (2016b) The tourism development-quality of life nexus in a small island destination. J Travel Res 55(1):79–94. https://doi.org/10.1177/0047287514532372

Smith A (1776; 2005) An inquiry into the nature and causes of the wealth of nations. The Pennsylvania State University

Song H, Witt S, Li G (2009) The advanced econometrics of tourism demand. Routledge, New York

Song H, Li G, Witt S, Fei B (2010) Tourism demand modelling and forecasting: how should demand be measured? Tour Econ 16(1):63–81

Strickland-Munro J, Allison H, Moore S (2009) Using resilience concepts to investigate the impacts of protected area tourism on communities. Ann Tour Res 32(2):499–519

UNWTO (2008) International recommendations for tourism statistics. United Nations Publication, New York

WTTC (2020) Travel & tourism: global economic impact & trends 2020. WTTC, London. https://wttc.org/Research/Economic-Impact. Accessed 1 Aug 2020

Chapter 4
Key Geographical Factors for Inbound and Domestic Tourism in Hokkaido

Soushi Suzuki and Peter Nijkamp

Abstract This chapter aims to provide an empirical contribution to trace key geographical factors for inbound and domestic tourism in Hokkaido Prefecture, Japan. In this study we use a set of relevant input and output data in 2015 for a set of 54 cities, towns and villages in Hokkaido Prefecture, so as to evaluate and compare their territorial tourism efficiency by means of Data Envelopment Analysis (DEA). We introduce a double-track efficiency approach, based on domestic tourism efficiency and inbound tourism efficiency. We find, on the one hand, that inbound efficiency scores and time distance from the tourism hub city (Sapporo) do not show a statistically significant outcome, while, on the contrary, domestic efficiency scores and time distance from the tourism hub city show statistically significant results. On the other hand, inbound efficiency scores and time distance from the nearest airport are statistically significant, whereas domestic efficiency scores and time distance from the nearest airport are not statistically significant. From the results, it is inferred that an improvement of both access to regional airports and of the frequency of international flights to regional airports may have a positive effect on inbound tourism and rural development in Hokkaido.

Keywords Data Envelopment Analysis (DEA) · Domestic tourism · Inbound tourism · Time distance from tourism hub city · Time distance from nearest airport

---

S. Suzuki (✉)
Hokkai-Gakuen University, Sapporo, Japan
e-mail: soushi-s@lst.hokkai-s-u.ac.jp

P. Nijkamp
Open University, Heerlen, The Netherlands

4.1 Introduction

Tourism is one of the key industries for realising sustainable development in developed countries, based on urban attractions, shopping functions, and historical and cultural resources and entertainment. On the other hand, rural areas are often assumed to be characterised by tranquillity, abundance of environmental resources, open space and traditional agricultural activities, though they usually suffer from poor access to advanced amenities, lack of shopping facilities, poor income-generating potential and closed social communities (see also de Noronha Vaz et al. 2013).

Nowadays, Japan is facing a depopulation and an ageing society; especially almost all rural areas in Japan have suffered from this cumbersome issue. The Government of Japan has positioned tourism as a major economic growth area and has implemented a Tourism Nation Promotion Basic Law (Japan Tourism Agency 2007). The Ministry of Land, Infrastructure, Transport and Tourism (MLIT) launched the MLIT Growth Strategy Council to examine and develop concrete measures for the ambitious tourism growth strategy. In addition, the Office of Tourism Nation Promotion was launched in 2009, establishing a unified government structure for making Japan an attractive destination for foreign tourists. Regarding the programme to attract 30 million international visitors to Japan, the Japanese government has worked to achieve the goal of 18 million visitors by 2016 through focussed communication about tourism attractions in priority markets as well as a joint public–private sector push for inbound travel promotion projects that support, among other things, the creation of attractive Japanese travel products (Japan Tourism Agency 2016).

These strategies have had the effect of increasing the number of international visitors; and indeed, the volume of international visitors to Japan was 31.19 million in 2018 (up 8.7% from the previous year), as is shown in Fig. 4.1.

Tourism is one of the main industries in Hokkaido Prefecture, and the total number of tourists visiting Hokkaido is ranked second among all prefectures in Japan. A time-series comparison of the number of international visitors to Hokkaido is shown in Fig. 4.2. From this figure, we notice that the number of international visitors (inbound) has kept on rising over the past years (2011–2017).

In 2013, the Government of Japan also established a policy to reform the ownership and management structure of all Japanese government-owned airports in a bid in order to reduce its large financial burden by involving private entities (see Miyoshi 2015). A consortium of private entities is now entering exclusive talks with the Land Ministry on operating New Chitose and six other airports in Hokkaido. The move is part of plans to privatise the operations of New Chitose, Wakkanai, Kushiro, Hakodate, Asahikawa, Obihiro and Memanbetsu airports. They are currently operated by the central or local government. The consortium could start operating the seven airports in stages starting in 2020 within a decade. It is expected that a 30-year contract for the airport operations will be concluded. The consortium is led by Hokkaido Airport Terminal Co., the operator of New Chitose Airport's terminal

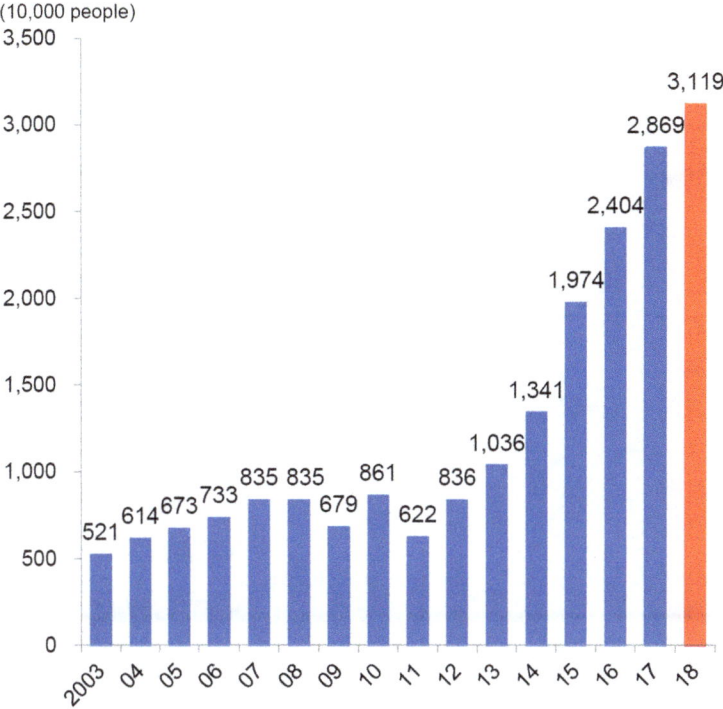

Fig. 4.1 The number of international visitors to Japan. (Source: White Paper on Tourism in Japan, Japan Tourism Agency 2019)

buildings. It also includes the Mitsubishi Estate Co., the Tokyu Corp., the state-affiliated Development Bank of Japan, the North Pacific Bank and the Hokkaido Electric Power Co. New Chitose Airport, which is close to Sapporo (the tourism hub city in Hokkaido), is financially stable, thanks to a strong demand from business travellers and tourists. The consortium aims to boost passenger numbers at the seven airports through demand-oriented measures such as developing sightseeing routes in Hokkaido (see The Japan Times 2019). A change in passenger numbers in these airports is predicted from 28.5 million to 45.8 million by 2049. The consortium has also made a business project may be in actual service of international flights to the six other airports.

These policy and project plans may have a positive effect for the development of rural tourism, but it has not yet been evaluated and has not yet demonstrated its effect in a quantitative way. The key question addressed in this chapter is, what key geographical factors (time distance from the tourism hub city, or time distance from the nearest airport) for inbound and domestic tourism can be related to a series of territorial background factors in these tourism areas (urban and rural)? In particular, the aim of our research is to identify possible differences in output–input ratios (i.e., in spatial efficiency or productivity patterns) between urban and rural tourism. It serves to answer the question, what type of tourism is mostly attracted by the

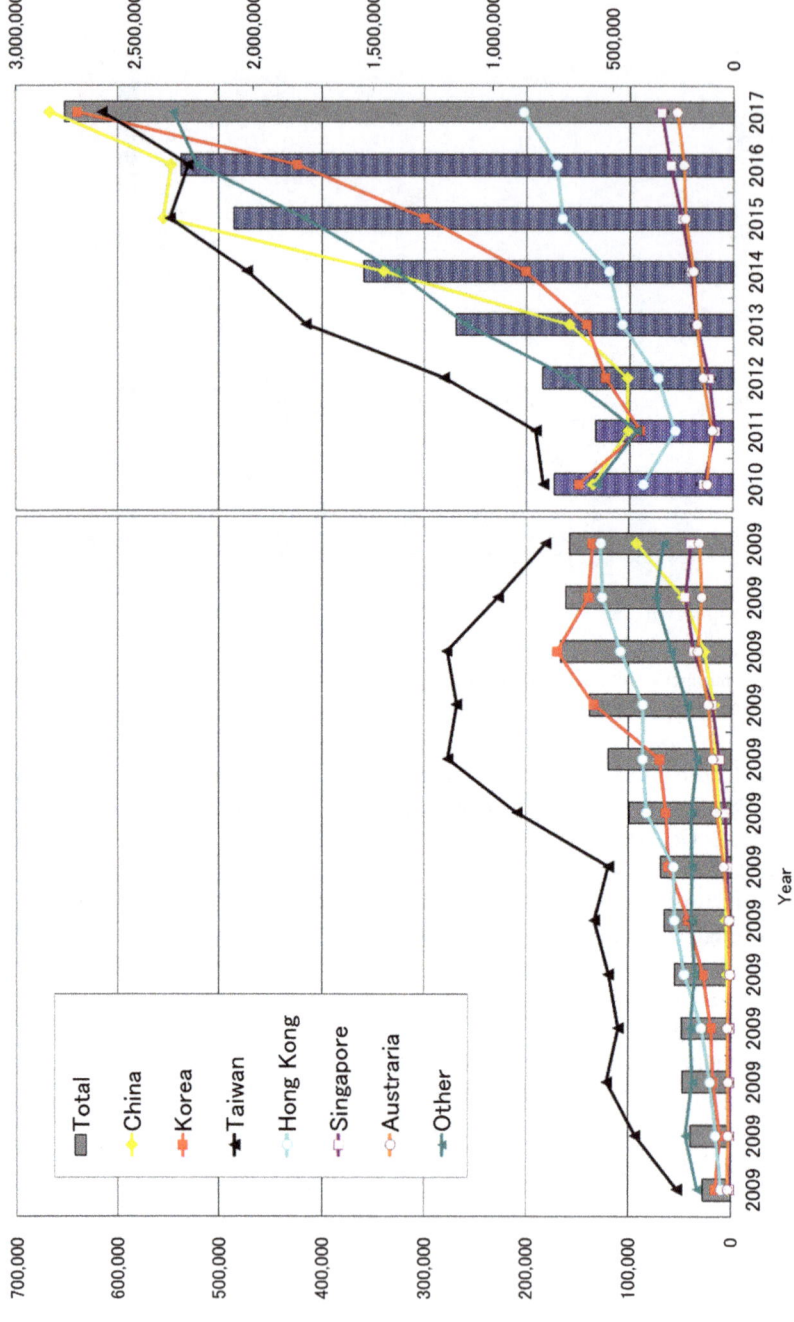

Fig. 4.2 Time-series comparison of the number of international visitors to Hokkaido. (Source: Hokkaido Government 2019)

tourist magnet functions of cities vs. rural areas? This research aim thus boils down to a comparative performance analysis of the efficiency of urban vs. rural tourism.

A standard tool which is used in decision theory and organisational management to judge the performance or efficiency among different agents or corporate organisations is Data Envelopment Analysis (DEA), initially proposed by Charnes et al. (1978). Over the past decades, this approach has become an established quantitative assessment method for comparative study and benchmark analysis in the evaluation literature. For instance, a decade ago Seiford (2005) mentioned that there were already at that time at least 2800 publications on DEA in various management and planning fields. Clearly, nowadays this number is much higher (for a review, see also Suzuki and Nijkamp 2017). In the meantime, the DEA methodology has expanded its scope towards other disciplines, such as economics, planning and environmental studies. Currently, in a tourism performance context, there are also several assessment studies that have applied DEA models to measure economic efficiency among cities, towns and villages, called Decision Making Units (DMUs) in the DEA jargon (see, e.g., Cracolici et al. 2008).

Various other introductions to DEA and its application to tourism efficiency rankings can be found in Suzuki et al. (2011), Shirouyehzad et al. (2012), Amin et al. (2013), Su (2013), Yin et al. (2015), Aissa and Goaied (2016), Huang et al. (2016), Zaman et al. (2016) and Chang et al. (2017). This large number of applied DEA studies already shows that an analysis of tourism efficiency in a competitive environment is not only an important but also an intriguing research topic in the regional science literature. DEA has demonstrated its great potential in providing a quantitative basis for comparative and benchmark studies in efficiency or productivity analysis (see for a broad overview, also Suzuki and Nijkamp 2017).

This chapter aims to provide an empirical contribution to trace key geographical factors for inbound and domestic tourism in Hokkaido Prefecture, Japan. In this study we use a set of relevant input and output data in 2015 for a set of 54 cities, towns and villages in Hokkaido Prefecture in order to evaluate and compare their territorial tourism efficiency by means of Data Envelopment Analysis (DEA). A critical element in the DEA application is the use of a Super-Efficiency DEA model to identify unambiguously the high performers among the efficient agents (i.e. tourist destinations).

The chapter is organised as follows. Section 4.2 briefly summarises the Super-Efficiency DEA methodology, while Sect. 4.3 presents the database and analytical framework employed in our study. Section 4.4 presents the evaluation results of the tourism efficiency analysis of cities, towns and villages in Hokkaido. Section 4.5 presents the results of a statistical significance test between key geographical factors (time distance from the tourism hub city and time distance from the nearest airport) and the performance score for each tourist destination. And finally, Sect. 4.6 draws some conclusions.

4.2   Outline of the SE-DEA Model

A broad introduction to DEA can be found in Suzuki and Nijkamp (2017). The unsatisfactory identification of efficient Decision-Making Units (DMUs) in a standard DEA model—where all efficient DMUs get the maximum score 1—has led to focussed research to discriminate between efficient DMUs, in order to arrive at a ranking—or even numerical rating—of these efficient DMUs, without affecting the results for the inefficient DMU's. In particular, Anderson and Petersen (1993) developed a radial Super-Efficiency (SE) DEA model. In general, an SE model aims to identify the relative importance of each individual efficient DMU, by assigning and measuring a score for its 'degree of influence', if this efficient DMU is omitted from the efficiency frontier (or production possibility set). If this elimination really matters (i.e. if the distance from this DMU to the remaining efficiency frontier is large), and thus the DMU concerned has a high degree of influence, and outperforms the other DMUs, it gets a high score (and is thus super-efficient). Thus, for each individual DMU, a new distance result is obtained, which leads to a new ranking—even a rating—of all original efficient DMUs.

These values are then used to rank the DMUs, and, consequently, efficient DMUs may then obtain an efficiency score above 1.000. The SE-DEA model may be suitable for our analysis in order to apply a statistical significance test between key geographical factors (time distance from the tourism hub city and time distance from the nearest airport) and the performance (DEA) score.

4.3   Database and Analytical Framework

An application of the SE-DEA model to the assessment of the performance of tourism will be presented here for the case of Hokkaido, Japan. In our empirical analysis on the efficiency of tourism destinations in terms of output–input ratios, we employ a set of relevant input and output data in 2015 for a set of 54 cities, towns and villages in Hokkaido Prefecture in order to evaluate and compare their tourism efficiency through the use of the SE-DEA method. The DMUs used in our analysis—essentially, the 54 tourist places—are listed in Fig. 4.3, which also contains relevant population information.

For our comparative analysis of these 54 municipalities, we consider two alternative efficiency viewpoints, and their related input and output items in the following way:

Efficiency viewpoint 1: Domestic tourism efficiency (4 inputs and 1 output).

Inputs (I):

(I1) Natural and ecological resources (attractiveness of sea, mountains, rivers, lakes, hot springs, leisure facilities and parks);

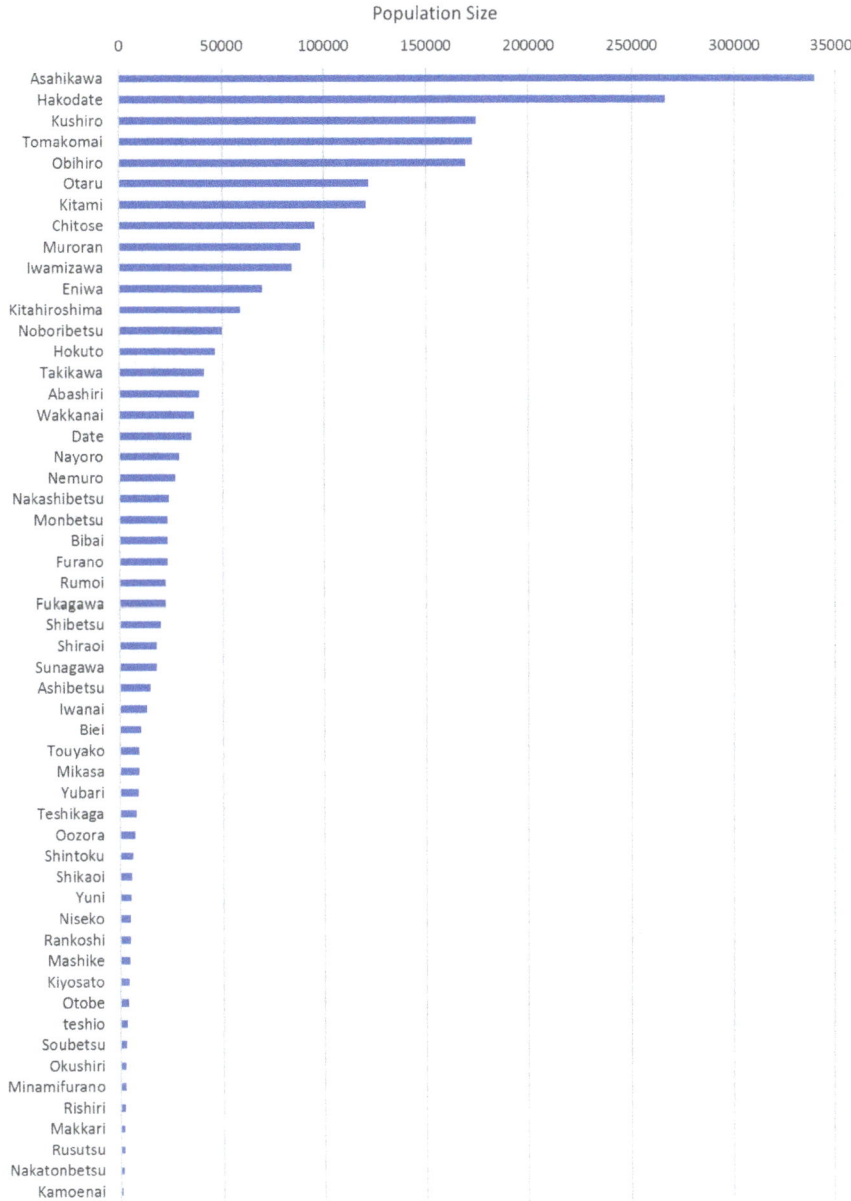

Fig. 4.3 List of population size of 54 cities, towns and villages in Hokkaido Prefecture by population size (Japan)

(I2) Historical and cultural resources (attractiveness of museums, historic public figures, streetscapes, historical architectures, traditional arts and festivals);
(I3) Shopping and food resources (attractiveness of shopping, souvenirs, quality of food and local foodstuff);
(I4) Hospitality and service resources (attractiveness of places, the hospitality from the locals, accommodation and accessible transportation systems).

Output (O):

(O1) Domestic visitor stays (total stays of Japanese travellers).

> *Efficiency viewpoint 2:* Inbound tourism efficiency (4 inputs and 1 output).
> Inputs (I): identical to the items mentioned at *efficiency viewpoint 1*;
> Output (O): (O2) Inbound visitor stays (total stays of foreign travellers).

All input datasets were collected from the official information provided by the Japanese 'Brand Research Institute: Regional Brand Research 2015'. The various attractiveness scores in the input system were calculated by using awareness survey results from a total of 29,046 respondents. All output datasets were collected from a report of the Tourist Bureau, Economics Department, Hokkaido Government: Report on Survey on Number of Tourists in Hokkaido, 2015.

The Hokkaido Prefecture has, in total, 179 municipalities, but the survey from the Regional Brand Research 2015 is limited to 55 municipalities, including Sapporo city. However, Sapporo city—with a total population of 1,953,784 and total tourism stays 13,652,800—was eliminated from our list of DMUs in order to avoid extreme and biased effects caused by statistical scale and score differences. Consequently, our study is limited to 54 DMUs.

4.4 Assessment of the Tourism Efficiency of Municipalities in Hokkaido

In this section, we present the comparative results of our SE-DEA model for 54 municipalities in Hokkaido Prefecture, for both domestic and foreign tourism, respectively.

4.4.1 Domestic Tourism Efficiency Evaluation Based on the SE-DEA Model

As mentioned above, we apply the SE-DEA model to the database with input and output items for the 54 DMUs in Hokkaido. The DEA results in terms of scores on domestic tourism efficiency for the 54 cities municipalities based on the SE-DEA model are presented in Fig. 4.4.

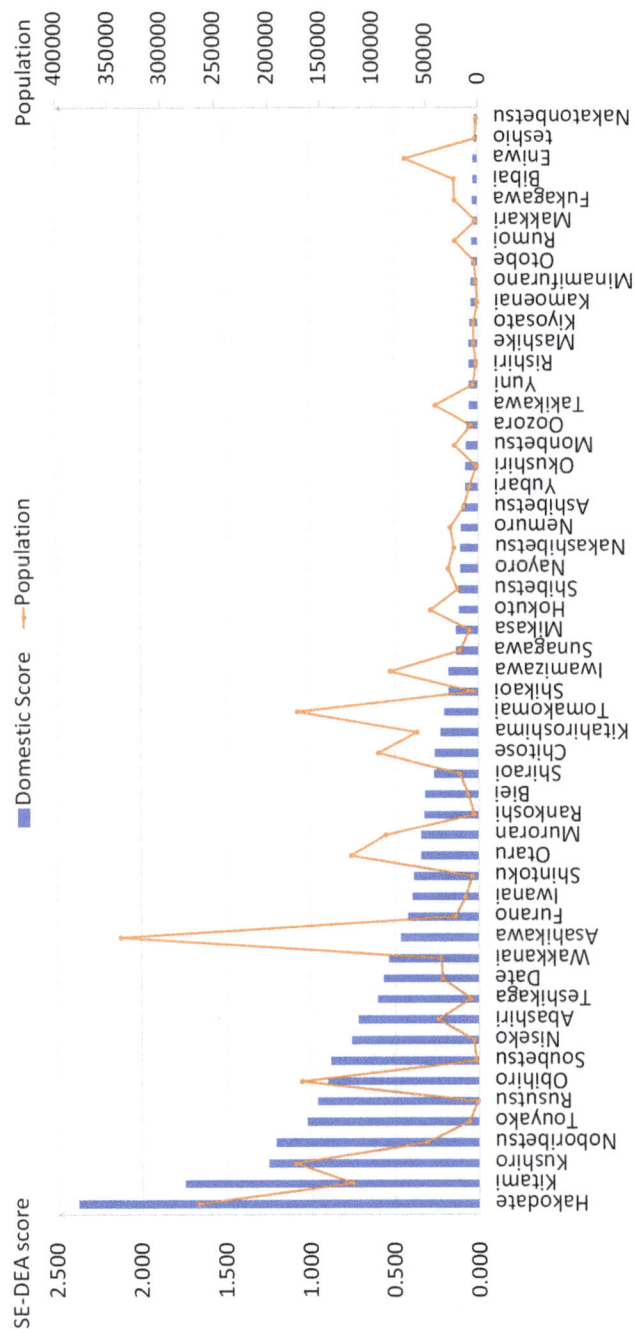

Fig. 4.4 Domestic tourism efficiency evaluation results for 54 municipalities in Hokkaido

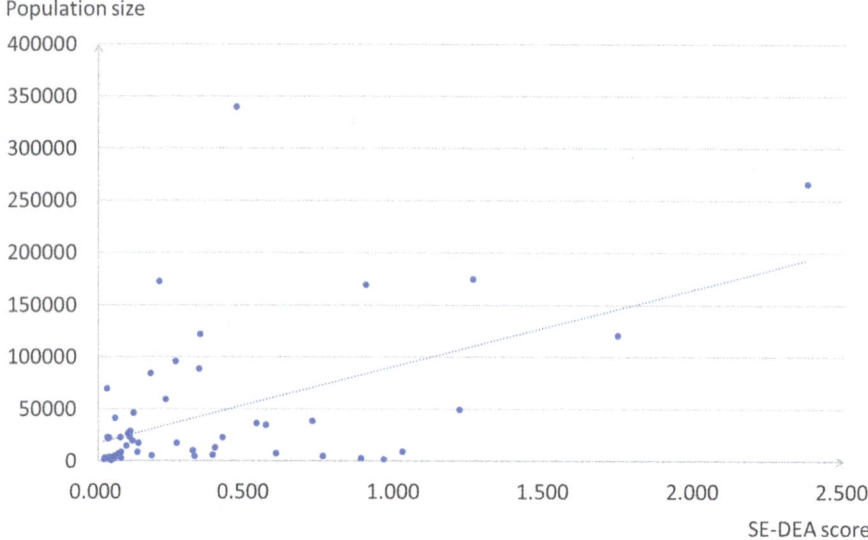

Fig. 4.5 Correlation between population size and domestic tourism efficiency

Table 4.1 Results of correlation analysis and statistical significance test

Item	Score
Correlation coefficient	0.509565046
t-value	3.148191478
p-value	0.00008305
Statistical significance	**

$*p < 0.05$
$**p < 0.01$

From Fig. 4.4, it can be seen that Hakodate, Kitami, Kushiro, Noboribetsu, and Toyako have an SE-DEA score above 1, and hence may be regarded as super-efficient tourism places. It can also be seen that the domestic tourism efficiency scores and population size tend to have a reasonable correlation. Therefore, we conducted a correlation analysis and a statistical significance test on these figures, as shown in Fig. 4.5 and Table 4.1.

From Table 4.1, it appears that the domestic tourism efficiency scores and the municipal population size have a clear statistical significance (with significance at the 1% level). From these results, we may infer that large cities and urban areas in Hokkaido have the ability to attract and effectively pull in more domestic tourists than its rural areas.

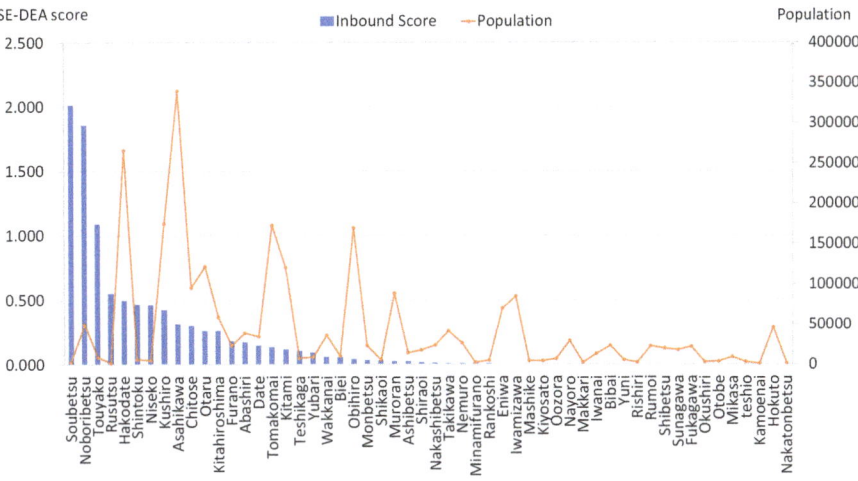

Fig. 4.6 Inbound tourism efficiency evaluation results for 54 municipalities in Hokkaido

4.4.2 *Inbound Tourism Efficiency Evaluation Based on the SE-DEA Model*

Next, we present the empirical findings for the efficiency of Inbound Tourism. These evaluation results for the 54 municipalities under consideration, based on the SE-DEA model, are given in Fig. 4.6.

Figure 4.6 shows that, seen from the perspective of foreign tourism, Soubetsu, Noboribetsu and Toyako may be regarded as super-efficient municipalities in Hokkaido. These municipalities are relatively low-populated places. Consequently, it seems plausible that the inbound tourism efficiency scores and the population size are likely to show a weak correlation. By applying a correlation analysis and performing a statistical significance test, as shown in Fig. 4.7 and Table 4.2, it turns out that indeed inbound tourism efficiency scores and population size are indeed not statistically significant. On the basis of these results, we may conclude that the towns in rural areas may have a better ability to be a magnet for inbound tourists.

4.5 Statistical Significance Test Between the Efficiency Score and Geographical Factors

In this chapter, we demonstrate a relationship between key geographical factors and the efficiency scores in order to derive policy implications for sustainable tourism and regional development. In this chapter we address especially two geographical

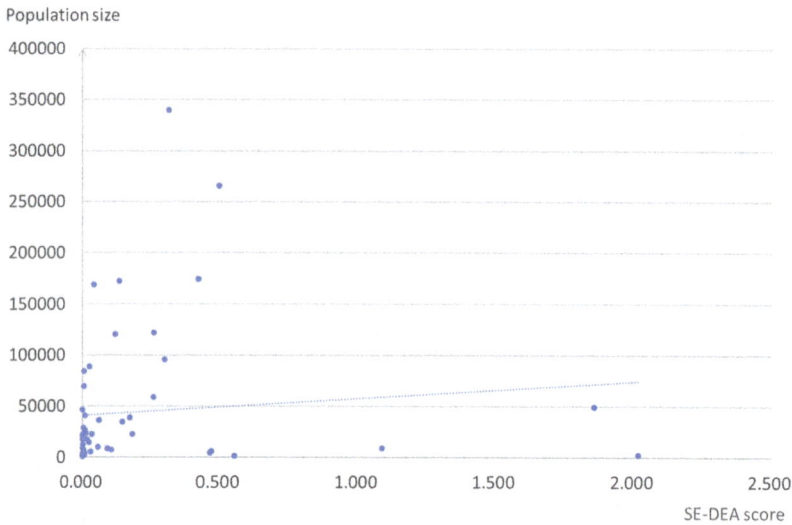

Fig. 4.7 Correlation between population size and inbound tourism efficiency in Hokkaido

Table 4.2 Results of correlation analysis and statistical significance test

Item	Score
Correlation coefficient	0.096230043
t-value	0.697160147
p-value	0.488807853
Statistical significance	Non-significant

$*p < 0.10$
$**p < 0.05$
$***p < 0.01$

factors, namely time distance from the tourism hub city and time distance from the nearest airport to each municipality.

4.5.1 Statistical Significance Test Between the Efficiency Score and Time Distance from the Tourism Hub City

Sapporo city—with a total population of 1,953,784 and total tourism stays of 13,652,800 in 2015—is a tourism hub city in Hokkaido. It is clear, therefore, that accessibility from Sapporo to each municipality is a key geographical factor for tourism. Based on this background, in this subsection we present a statistically significant test result concerning the relationship between efficiency score and time distance from Sapporo city. This geographical information is presented in Fig. 4.8.

In this analysis, we measured time distance (min) between Sapporo and each municipality by means of a Google map function, and we assumed that all trips were

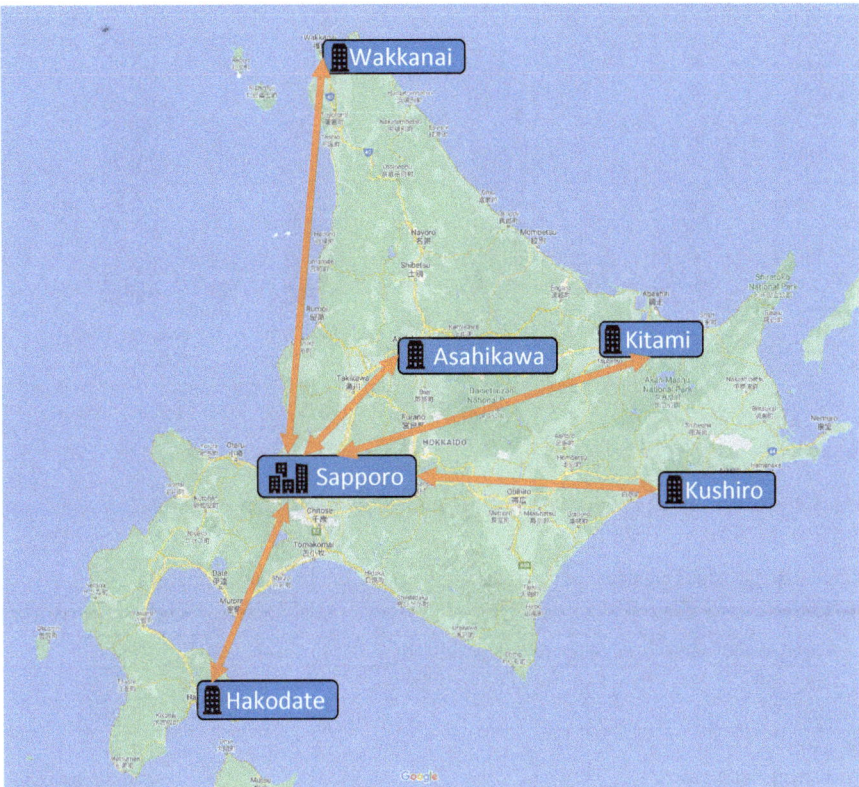

Fig. 4.8 A diagrammatic representation of time distance from the tourism hub city (Sapporo) to each municipality. (Source: Google map, and edited by author)

made by car travel. Then we distinguished 2 DMU groups: namely 27 DMUs with high efficiency-score DMUs and 27 with low efficiency-scores DMUs. A comparison for average time distance from Sapporo city to each DMU group for each inbound and domestic result is shown in Fig. 4.9. While also the results of a statistical significance tests (Student's t-test) of these average time distances are shown in Table 4.3 (Inbound) and Table 4.4 (Domestic).

From these results, we notice that inbound efficiency scores and time distance from the tourism hub city (Sapporo) are not statistically significant, while on the contrary, domestic efficiency scores and time distance from the tourism hub city have weak statistical significance. Based on these results, it is inferred that an improvement of access to Sapporo city to each municipality may have a positive effect on domestic tourism.

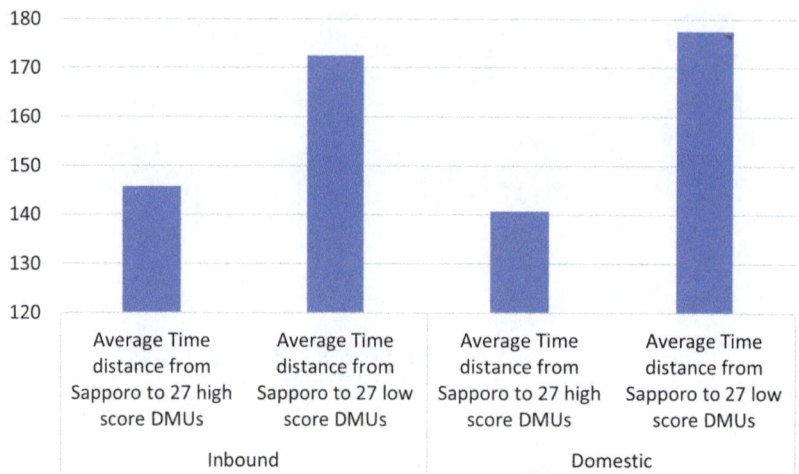

Fig. 4.9 Comparison of average time distance from Sapporo city to each DMU groups

Table 4.3 Statistical significance test (Student's *t*-test) result for inbound tourism

Item	Score
Average time distance from Sapporo city (min) for the 27 high-score DMUs	145.778
Average time distance from Sapporo city (min) for the 27 low-score DMUs	172.481
t-value	0.899
p-value	0.187
Statistical significance	Non-significant

*$p < 0.10$
**$p < 0.05$
***$p < 0.01$

Table 4.4 Statistical significance test (Student's *t*-test) result for domestic tourism

Item	Score
Average time distance from Sapporo city (min) for the 27 high-score DMUs	140.704
Average time distance from Sapporo city (min) for the 27 low-score DMUs	177.556
t-value	1.405
p-value	0.083
Statistical significance	*

*$p < 0.10$
**$p < 0.05$
***$p < 0.01$

Fig. 4.10 An image of the time distance from the nearest airport to each municipality. (Source: Hokkaido Regional Development Bureau, and edited by author)

4.5.2   Statistically Significance Test Between the Efficiency Score and the Time Distance from the Nearest Airport

Accessibility from the nearest airport to a municipality is also a key geographical factor for tourism. Based on this background information, in this subsection we present statistical significance test results between the efficiency score and the time distance from a nearest airport. A diagrammatic representation of this geographical information is shown in Fig. 4.10.

In this analysis, we measured time distance (in min) from the nearest airport to each municipality office by means of the Google map information, and we assumed that all trips were again made by car. Next, we distinguished again 2 DMU groups with 27 DMUs with high efficiency score DMUs and 27 low-efficiency score DMUs. By comparing the average time distance for all inbound and domestic results, we obtain Fig. 4.11. The results of a statistical significance t-test of these average time distances are shown in Table 4.5 (Inbound) and Table 4.6 (Domestic).

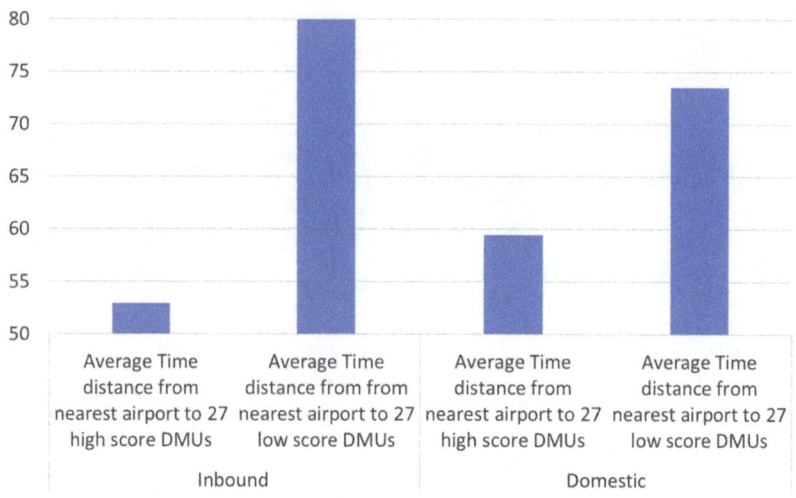

Fig. 4.11 Comparison of the average time distance from the nearest airport to each DMU group

Table 4.5 Statistical significance test (Student's *t*-test) result for inbound tourism

Item	Score
Average time distance from nearest airport (min) for high-score DMUs	52.963
Average time distance from nearest airport (min) for low-score DMUs	80.000
t-value	2.364
p-value	0.011
Statistical significance	**

$*p < 0.10$
$**p < 0.05$
$***p < 0.01$

Table 4.6 Statistical significance test (Student's *t*-test) result for domestic tourism

Item	Score
Average time distance from nearest airport (min) for high-score DMUs	59.444
Average time distance from nearest airport (min) for low-score DMUs	73.519
t-value	1.185
p-value	0.121
Statistical significance	Non-significant

$*p < 0.10$
$**p < 0.05$
$***p < 0.01$

From these results, we derive that inbound efficiency scores and time distance from the nearest airport are statistically significant, whereas domestic efficiency scores and time distance from the nearest airport are not a statistically significant. Based on these results, it is also inferred that an improvement of access to regional airports and in the frequency of international flights to regional airports may have a positive effect on inbound tourism and rural development.

4.6 Concluding Remarks

In this chapter, we have addressed the question whether the behaviour of domestic and inbound tourism shows a distinct pattern with regard to urban or rural areas, and if so, why. Our SE-DEA approach has found not only a clear ranking of tourist destinations in terms of their urban and rural character, but has also identified a clear correlation between these rankings and the volumes of domestic and inbound tourists. Clearly, our findings refer in particular to the case of Hokkaido, Japan. Thus, there is much scope for an extension of our research towards other areas, based on individual appreciation data related to other tourism places.

From our analysis results, we may infer that rural areas may have a better ability to be a magnet for inbound tourists. We also found that inbound efficiency scores and time distance from the nearest airport are statistically significant. Based on these results, it is inferred that an improvement of both access to regional airports and the frequency service of international flights to regional airports may have a positive effect on inbound tourism and rural development.

These findings have important policy implications. If the beauty of rural or natural areas is a greater attraction force for foreign visitors than for domestic visitors—a result also found in the study by Giaoutzi and Nijkamp (1993)—, the marketing of tourism in rural areas should more explicitly address international markets, based on dedicated marketing strategies in promising countries of origin.

References

Aissa SB, Goaied M (2016) Determinants of Tunisian hotel profitability: the role of managerial efficiency. Tour Manag 52:478–487

Amin M, Yahya Z, Ismayatim WFA, Nasharuddin SZ, Kassim E (2013) Service quality dimension and customer satisfaction: an empirical study in the Malaysian hotel industry. Serv Mark Q 34 (2):115–125

Anderson P, Petersen N (1993) A procedure for ranking efficient units in data envelopment analysis. Manag Sci 39:1261–1264

Chang YT, Lee S, Park HK (2017) Efficiency analysis of major cruise lines. Tour Manag 58:78–88

Charnes A, Cooper WW, Rhodes E (1978) Measuring the efficiency of decision-making units. Eur J Oper Res 2:429–444

Cracolici MF, Nijkamp P, Rietveld P (2008) Assessment of tourism competitiveness by analyzing destination efficiency. Tour Econ 14(2):325–342

Giaoutzi M, Nijkamp P (eds) (1993) Decision support models for regional sustainable development. Ashgate, Aldershot

Hokkaido Government (2019) Time-series comparison of the number of visitors in Hokkaido. http://www.pref.hokkaido.lg.jp/kz/kkd/irikominosuii.htm

Huang SW, Kuo HF, Hsieh HI, Chen TH (2016) Environmental efficiency evaluation of coastal tourism development in Taiwan. Int J Environ Sci Dev 7(2):145–150

Japan Tourism Agency (2007) Tourism nation promotion basic law. http://www.mlit.go.jp/kankocho/en/kankorikkoku/kihonhou.html

Japan Tourism Agency (2016) T inbound travel promotion project (Visit Japan Project). http://www.mlit.go.jp/kankocho/en/shisaku/kokusai/vjc.html

Japan Tourism Agency (2019) White paper on tourism in Japan. http://www.mlit.go.jp/kankocho/en/siryou/content/001312296.pdf

Miyoshi C (2015) Airport privatisation in Japan: unleashing air transport liberalisation? J Airport Manag 9(3):210–222

de Noronha Vaz T, van Leeuwen E, Nijkamp P (eds) (2013) Towns in a rural world. Ashgate, Aldershot

Seiford L (2005) A cyber-bibliography for data envelopment analysis (1978–2005). CD-ROM in introduction to data envelopment analysis and its uses. Springer Science+Business Media, New York

Shirouyehzad H, Lotfi FH, Shahin A, Aryanezhad MB, Dabestani R (2012) A DEA approach for comparative analysis of service quality dimensions with a case study in hotel industry. Int J Serv Oper Manag 12(3):289–308

Su CS (2013) An importance-performance analysis of dining attributes: a comparison of individual and packaged tourists in Taiwan. Asia Pac J Tour Res 18(6):573–597

Suzuki S, Nijkamp P (2017) Regional performance measurement and improvement: new developments and applications of data envelopment analysis, new frontiers in regional science: Asian perspectives. Springer, Berlin

Suzuki S, Nijkamp P, Rietveld P (2011) Regional efficiency improvement by means of data envelopment analysis through Euclidean distance minimization including fixed input factors: an application to tourist regions in Italy. Pap Reg Sci 90(1):67–89

The Japan Times (2019) Consortium to enter talks with Japan Land Ministry on Privately Operating Seven Hokkaido Airports. https://www.japantimes.co.jp/news/2019/07/04/business/consortium-enter-talks-land-ministry-privately-operating-seven-hokkaido-airports/

Yin P, Tsai H, Wu J (2015) A Hotel Life Cycle Model based on bootstrap DEA efficiency: the case of international tourist hotels in Taipei. Int J Contemp Hosp Manag 27(5):918–937

Zaman M, Botti L, Thanh TV (2016) Does managerial efficiency relate to customer satisfaction? The case of Parisian Boutique Hotels. International Journal of Culture, Touri

Part III
Tourism and Historical Heritage

Chapter 5
Tourism, Leisure and Cultural Heritage: The Challenge of Participatory Planning and Design

Gert-Jan Burgers

Abstract This chapter is about democratization and citizen participation in the management of cultural heritage. Although heritage is often perceived as a domain of experts, it is key to the daily lives of citizens. Increasingly, as in nearly all sectors of society, citizens demand a voice in the definition and management of heritage, and in the development of planning alternatives and design solutions, amongst others with regard to tourism, leisure and recreation. Here, heritage planning meets a UN sustainable development goal, that of inclusive and equal social justice. Government agencies, heritage professionals and spatial planners are already beginning to open up to the public, aiming to increase inclusiveness, and heritage tourism and recreation is accessible to larger sections of society than ever. However, there is much debate, but little research, on current concepts, tools and procedures for democratization in the access to and definition, appropriation, management and planning of heritage. It is the explicit aim of the EU-funded Project Heriland to explore such concepts, tools and procedures in a series of laboratory contexts throughout Europe, both urban and rural. The Heriland Project is an International Training Network (ITN), funded through the EU Horizon2020 Marie Curie Action (GA 813883; 2019–2023). It is a collaboration of VU University Amsterdam, University of Newcastle Upon Tyne, Goteborgs Universitet, Universita' degli Studi Roma Tre, Technische Universiteit Delft, the Bezalel Academy of Arts and Design and 16 associated partners from all over Europe. In this chap. I present one of these labs, that of the so-called *Ecomuseo della* Via *Appia*, in a rural context in the southern Italian Apulia region.

Keywords Cultural heritage · Landscape · Democratization · Ecomuseum · Citizen co-creation

G.-J. Burgers (✉)
VU University Amsterdam, Amsterdam, Netherlands
e-mail: g.l.m.burgers@vu.nl

5.1 Introduction

Cultural heritage is one of the most highly valued assets in modern tourism, leisure and recreation. Most countries have a lengthy and successful history of conserving heritage and capitalizing on it culturally and economically. Throughout the twentieth century, particularly since the 1960s, great progress has been made in creating structures and promulgating principles to guide heritage and landscape conservation, often in conjunction with the international community through UNESCO (e.g. Emerick 2014; Smith 2006; Gibson and Pendlebury 2009; Fairclough et al. 2008; Janssen et al. 2017). As the twenty-first century proceeds, however, it is becoming increasingly clear that a further paradigm shift is required. There are new far-reaching drivers for change, including rising and moving populations, greater connections through the digital world between communities, mass tourism and the sustainability movement. Also, democratization increasingly questions the role of professionals in guiding developments. The significant steps forward made in heritage theory, aims and practice are no longer sufficient. Confronted with such a fast-changing context, heritage management needs to become more proactive. More powerful ideas, tools and training are needed to ensure that interdisciplinary, research-based heritage and landscape management is positively integrated with business activity, with city and rural development, and with democratic participation in decision making.

This chapter is about democratization and citizen participation in heritage management. Although cultural heritage is often perceived as a domain of experts, it is key to the daily lives of citizens. Increasingly, as in nearly all sectors of society, citizens demand a voice in the definition and management of heritage, and in the development of planning alternatives and design solutions, amongst others with regard to tourism, leisure and recreation (e.g. Smith 2006; Harvey 2006; Harrison 2010, 2013; Neal 2015). Here, heritage planning meets a UN sustainable development goal, that of inclusive and equal social justice. Government agencies, heritage professionals and spatial planners are already beginning to open up to the public, aiming to increase inclusiveness, and heritage tourism and recreation is accessible to larger sections of society than ever. However, there is much debate, but little research, on current concepts, tools and procedures for democratization in the access to and definition, appropriation, management and planning of heritage.

Such research is central to a new European Training Network, which I am happy to coordinate, and which carries the acronym *Heriland*, which stands for Cultural *Heri*tage and the planning of European *Land*scapes (Fig. 5.1). It is a collaboration of six universities, VU University Amsterdam, University of Newcastle Upon Tyne, Goteborgs Universitet, Universita' degli Studi Roma Tre, Technische Universiteit Delft and the Bezalel Academy of Arts and Design, and 16 associated partners from all over Europe. The overriding aim of Heriland is the empowerment of a new generation of academics, policy makers, practitioners, professionals and entrepreneurs. This new generation must devise and guide transdisciplinary planning and design strategies for regenerating European heritage and landscape, foster social

Fig. 5.1 The Heriland Project. © Bert Brouwenstijn, VU University Amsterdam

inclusiveness and create socially, economically and environmentally sustainable future landscapes. One of the major objectives of the project is to devise an innovative set of concepts, techniques and skills for promoting democratic, co-creative approaches. This is being done in a series of laboratory contexts throughout Europe, both urban and rural.[1] In this chap. I present one of these labs,

---

[1]The Heriland Project is an International Training Network (ITN), funded through the EU Horizon2020 Marie Curie Action (GA 813883; 2019–2023). See especially www.heriland.eu

that of the so-called E*comuseo della* Via *Appia*, in a rural context in the southern Italian Apulia region.

5.2   The Ecomuseum of the Via Appia (EVA)[2]

The *Ecomuseo della* Via *Appia*, or EVA in short, is a recent initiative of citizens of the Italian municipalities of Latiano and Mesagne (Brindisi, Italy), implemented in collaboration with *VU University Amsterdam*.[3] Its aim is to enhance citizen participation in the definition of and care for the cultural heritage of the local district, located along the final stretch of the famous ancient road known as the Via Appia.

The installation of the ecomuseum is intimately linked to the gradual reshaping of the cultural landscape between the municipalities of Mesagne and Latiano, and in particular to the exploration and valorization of the archaeological site of Muro Tenente. The ruins of this 50 ha site represent more than 3000 years of local history and include monumental fortifications, cemeteries and sacred buildings, buried under extensive vineyards, olive groves and uncultivated fields. Since the 1970s, they have progressively been unearthed by teams of the regional heritage board and *VU University Amsterdam* (excavations *in concessione* of the Italian Ministry of Culture, *MIBACT*) (Figs. 5.2 and 5.3. Alberda et al. 1999, Burgers 1998, Burgers and Napolitano 2010, Burgers and Yntema 1999). Both institutions have also put much energy into the valorization of the site, in collaboration with the municipalities of Mesagne and Latiano, local interest groups, and the *Università del Salento*. Together, they also proposed the creation of an archaeological landscape park in the area, the so-called Parco Archeologico di Muro Tenente. (Fig. 5.4).[4]

In the process of formation of the park, much value has been attached to recent arguments against the establishment of such parks throughout Europe. As a matter of fact, at the start of the third millennium, the archaeological park has become a widespread phenomenon in Europe and beyond. However, it has met with considerable criticism, particularly from archaeologists (e.g. Hodder 1992; Bender 1992; Kolen 1995). The opponents object to the fact that the parks are commonly part of a top-down approach. A related argument against establishing such parks is that the

---

[2]Part of this section is adapted from Opmeer et al. (2019). My sincere thanks go to my co-authors in that paper, i.e. Mark Opmeer, Christian Napolitano, Ilaria Ricci and Rosanne Bruinsma. Cf. Burgers et al. (2018).

[3]Our special thanks go to the *Cooperativa Impact* and its members, especially Christian Napolitano, Arturo Clavica, Ilaria Ricci, Margherita De Matteis, Sara De Girolamo, Patricia Caprino.

[4]Many thanks go to our colleagues of the *Soprintendenza archeologia, belle arti e paesaggio per le province di Brindisi, Lecce e Taranto* (especially the Soprintendente Arch. Maria Piccarreta, d.ssa Assunta Cocchiaro and d.ssa Annalisa Biffino), the *Universita' degli Studi del Salento* (especially proff. Francesco D'Andria, Grazia Semeraro and Francesco Baratti) and the municipalities of Mesagne (the mayor dott. Antonio Matarrelli, Domenico Stella and d.ssa Concetta Franco) and Latiano (the mayor Cosimo Maiorano, d.ssa Margherita Rubino).

Fig. 5.2 The Muro Tenente Project. © Bert Brouwenstijn, VU University Amsterdam

Fig. 5.3 Artist reconstruction of the ancient town buried at Muro Tenente. © Mikko Kriek. VU University Amsterdam

model is European and that they would reflect a mainstream view of European or world history, at the expense of local perceptions of history and landscape. The main opponents argue that such parks imprison history and expel it from daily life and experience, since people visit these places primarily for the purpose of recreation (e.g. Hodder 1992, Bender 1992, Kolen 1995). The argument goes that such parks are alien to the local context, since, from a local perspective, the past is generally an integral part of everyday life, made up of personal biographies and family histories rather than notions of world history. People literally live in historical environments, making them productive. A good example of this is the abundance of ancient cave sanctuaries in Apulia which are often used for storage or for keeping animals. The parks' opponents argue that it is these local perceptions that we have to respect. They even advocate going a step further and letting local communities manage their own landscapes and shape their own histories, even at the risk of destroying the archaeological record.

To a certain extent, I agree with such views, but let me start by voicing my misgivings. First, I hold that this view is essentially dualistic, dividing up the world into outside policy makers and professional archaeologists on the one hand and inside local communities on the other. However, look a little closer and it is easy to recognize a myriad of grey areas between and within both groups. Second, the outsiders are presented as negative and local perceptions as inherently better and therefore preferable. In my opinion, this is a romantic view which idealizes a pre-rational state of local society. The connotation is that locals are "wild" and "pure" and have to be protected against colonizing heritage specialists and archaeologists.

I would argue that this image needs to be refined; local communities are also part of modern society and continuously evolving. Their present perceptions of the world may even be more rational and economically motivated than those of outside policy

Fig. 5.4 Master plan of the archaeological park of Muro Tenente. © Francesco Baratti

Fig. 5.5 Summer school at the excavations at Muro tenente. © Susanne Strijbis, VU University Amsterdam

makers. What is more, their past perceptions of history and landscape may be outspokenly negative, and in many cases their wish to shake off these perceptions is strong. In southern Italy in particular, archaeological sites have long been depicted as demonic, and the countryside in general has traditionally been associated with such negative phenomena as the tolls of forced labour in the service of large estate owners. It should therefore come as no surprise to learn that the initiative of setting up archaeological parks comes from within the local communities themselves. Indeed, many experience the discovery of a new history and a new heritage in the context of a more general emancipation from traditional society. Furthermore, they realize that the traditional perception and use of the historical landscape is fatal to its preservation.

It should be clear that, together with the local communities, I too favour conservation. But I also agree with the opponents that we should beware of adopting an international blueprint of a fossilized archaeological park. In every single case, I think it is wisest to integrate parks into their local context and, within limits, to respect such matters as the local use of the landscape and local perceptions of history. This indeed reflects our approach in Apulia. From the very start, therefore, the local communities have been actively involved in this endeavour, through school visits, guided tours, public lectures and training sessions (Fig. 5.5), among others, but also through exhibitions and open air festivals (Fig. 5.6), in which their own heritage perspectives and values were promoted. They were also involved in the planning of the park.

Fig. 5.6 Muro Tenente festival 2020. © Christian Napolitano

Together with citizens and town councils, we even took local input a step further, exploring new, more democratic heritage practices and policies. It is at this point that the concept of ecomuseum was put forward. The concept is rooted in an international trend, which started in the first half of the previous century, when "radical" ideas started to emerge that questioned the authoritative roles of museums (Davis 2008). This was a slow process, which culminated in what is nowadays called the "new museology" in the 1970s. A landmark was the Round Table on the Development and the Role of Museums in the Contemporary World, organized by UNESCO and ICOM in 1972 (Guido 1973; Davis 2008). The resolutions made here were geared towards creating greater societal responsibility for museums, in which the local community became the prime stakeholder. Accordingly, instead of targeting objects and sites and "freezing" them in physical museum buildings, the focus was placed on communities and the way they define heritage in the context of their immediate living environment. A prime example of this new approach is ecomuseology, which promotes democratic participation in the interpretation of local history and the management of local heritage (e.g. Howard 2002; Van Mensch 2005; Davis 2008; Crooke 2010).

Ecomuseums are not buildings with collections, but refer to a specific living environment and its inhabitants (e.g. villages, urban neighbourhoods or industrial peripheries). They commonly aim to strengthen community bonds by engaging communities in the management of their "own" heritage, thus strengthening links between the present and the past. Initially (1970s–2000s), ecomuseums were conceived from an essentialist perspective as relatively closed socio-spatial entities. They were established especially in rural communities, because of their perceived social stability, traditionalism and shared, homogeneous history and heritage. However, ecomuseums, as other new museologies, are now increasingly adopting pluralist and constructivist attitudes towards heritage management. The latter approach is also key to the *Ecomuseo della* Via *Appia*, with its aim of giving a voice to the multiple stakeholder groups interested in the cultural heritage of the Brindisi district, from civil society institutions to individual citizens and local entrepreneurs. The main objectives of EVA are social cohesion, development of the tangible and intangible heritage, regional and local economic development and improvement of the quality of the landscape (Baratti 2012: 22). Of these objectives, enhancing social cohesion was considered to be key, as without it, the whole ecomuseum could be considered as being deprived of its fundamental meaning, making it more difficult to catalyse the economic growth of the heritage (Borrelli and Davis 2012: 42–43).

In order to fulfil these aims, the following strategies were used: knowledge sharing, storytelling and education (Fig. 5.7), active community participation, protection and conservation, rediscovery of traditional jobs, sustainable tourism, marketing and promotion of the area, and eco-sustainable management (Davis 1999; Corsane and Holleman 1993; Maggi 2006). The strategies selected were discussed and expanded together with local citizens and stakeholders. Only after a process of active participation is it possible to develop concrete proposals for an efficient, long-term project, which is not just imposed on the area by external actors (Borrelli and Davis 2012: 32). The targets, divided into four main categories (social, patrimonial,

Fig. 5.7 Knowledge sharing and storytelling as part of the activities of the Ecomuseo della Via Appia. © Christian Napolitano

economical and territorial), were linked to the strategies inside a *value tree*. This is a flux diagram used by the group in order to increase its understanding of the motivations and the aims of the project and establish a hierarchy among them, after analysing each of them in relation to the strategies and the categories. This, in turn, made it possible to define several activities to increase social cohesion; throughout the years we have organized a long series of activities, from discussion platforms to interactive workshops, from music festivals to theatre productions, and from bicycle tours to natural, cultural and Eno-gastronomic guided tours. Systematic interviews and questionnaires have been organized amongst others to investigate the degree of uniformity or heterogeneity between the voices, perceptions, needs and wishes of various social groups.

Of particular interest was a series of interactive workshops aimed at creating the elaborate *Mappa della Comunità* (Fig. 5.8), which charts the progress of the formation of local and regional development plans, notably the regional *Piano Paesaggistico Territoriale Regionale della Puglia*. The *Mappa della Comunità* is considered as a tool for spurring the growth of the "awareness of a place" by having citizens contribute to constructing the representation of heritage, territorial and landscape values.[5] By doing so, citizens are supported by facilitators, artists and local historians, ultimately producing a map that portrays "a part of the landscape as it is perceived by local residents".[6] The most important aims are: mapping the local

---

[5]*Art. 13, comma 1, dell'elaborato 2 "Norme tecniche di attuazione"*, attached to the proposal of *Piano Paesaggistico Territoriale della Regione Puglia (PPTR)*. Many thanks go to Mr. Enzo Camassa for the end production of the Latiano map.

[6]Art. 1 European Landscape Convention, 2000.

Fig. 5.8 The *Mappa della Comunita'* of Latiano. ©Enzo Camassa

perception of the landscape, gaining an understanding of the territory as a representation of the history of the area as preserved in individual and collective memory, and
communicating a reading of the values of the landscape to local residents, but, above
all, promoting a "community pact", which binds citizens, operators and institutions
to the landscape (Baratti 2012: 76–93).

5.3   Conclusion

There are strong indicators demonstrating that the above-discussed participative
approach has been successful. Thus, local groups of various ages and social back-
grounds are now visiting the ecomuseum sites on their own initiative, although there
is often nothing more to see than pottery shards and abandoned fields. They have
even started organizing open-air festivals or to develop bicycle tours and athletics
games in the area (Fig. 5.9). This is a remarkable development because until recently
these sites were regarded as no-go areas, spurned for their negative heathen associ-
ations far from the mother church in town. Most local people did not experience
these sites as part of their heritage at all. Now, local sensibility is awakening. The
communities are slowly starting to identify with the historical landscapes in a
positive way and are appropriating them as their legitimate heritage. Consequently,
they are also taking up responsibility for the care of these landscapes, together with
the local authorities which formally propose and coordinate the projects. A clear sign

Fig. 5.9   Athletics games at Muro Tenente. © ASD Olimpo Latiano

Fig. 5.10   Inauguration of a new visitor area at Muro Tenente, with reconstructed ancient dwelling. © Quimesagne.it

of this new sensibility is also that the town councils involved, those of the municipalities of Mesagne and Latiano, have applied for European Union funding in order to institutionalize and expand the landscape park of the ecomuseum (Fig. 5.10). This funding has recently been granted, within the context of a much larger heritage programme coordinated by the Regione Puglia. Such projects now even constitute an issue in election campaigns and can therefore be seen as reflecting the sensibility of wider sections of the local population, without whose support they would never have been proposed in the first place.

References

Alberda K, van Burgers GJ, Burgers H, Karel D, Yntema D (1999) Muro Tenente. Centro messapico nel territorio di Mesagne, a cura di A. Nitti, Mesagne, Italy
Baratti F (2012) Ecomusei, paesaggi e comunità. Esperienze, progetti e ricerche nel, Salento, Sogein, Lecce
Bender B (1992) Theorising landscapes, and the prehistoric landscapes of Stonehenge. Man (N.S.) 27(4):735–755
Borrelli N, Davis P (2012) How culture shapes nature: reflections on ecomuseum practices. Nat Cult 7(1):31–47
Burgers GJ (1998) Constructing Messapian landscapes: settlement dynamics, social organization and culture contact in the margins of Graeco-Roman Italy. Gieben Publishers, Amsterdam
Burgers GJ, Napolitano C (2010) L'insediamento messapico di Muro Tenente. Scavi e ricerche 1998-2009, Mesagne
Burgers GJ, Yntema DG (1999) The settlement of Muro Tenente. Third interim report. Bulletin Antieke Beschaving 74(2):111–132
Burgers GJ, Napolitano C, Piccarreta M (2018) Parco dei Messapi di Muro Tenente: un progetto di sviluppo sostenibile. Bollettino di Archeologia On Line. Direzione Generale Archeologia, Belle Arti e Paesaggio IX 2–3:1–19
Corsane G, Holleman W (1993) Ecomuseums: a brief evaluation. In: De Jong R (ed) Museums and the environment. Southern Africa Museums Association, Pretoria, pp 111–125
Crooke E (2010) The politics of community heritage: motivations, authority and control. Int J Herit Stud 16(2):16–29
Davis P (1999) Ecomuseums: a sense of place. A&C Black, London
Davis P (2008) New museologies and the ecomuseum. In: Graham B, Howard P (eds) The Ashgate research reader in heritage and identity. Ashgate, Aldershot, pp 397–414
Emerick K (2014) Conserving and managing ancient monuments: heritage, democracy, and inclusion. Heritage Matters Series 14. Boydell Press, Woodbridge
Fairclough G, Harrison R, Jameson J Jr, Schofield J (eds) (2008) A heritage reader. Routledge, Abingdon
Gibson L, Pendlebury J (eds) (2009) Valuing historic environments. Ashgate, Farnham
Guido HF (1973) UNESCO Regional seminar: Round table on the development and the role of museums in the contemporary world. UNESCO document SHC.72/CONF.28/4
Harrison R (2010) Heritage as social action. In: West S (ed) Understanding heritage in practice. Manchester University Press, Manchester, pp 240–276
Harrison R (2013) Heritage: critical approaches. Routledge, Abingdon
Harvey D (2006) The right to the city. In: Scholar R (ed) Divided cities: the Oxford Amnesty lectures 2003. Oxford University Press, New York, pp 83–103
Hodder I (1992) Theory and practice in archaeology. Taylor & Francis, London
Howard P (2002) The eco-museum: innovation that risks the future. Int J Herit Stud 8(1):63–72

Janssen J, Luiten E, Renes H, Stegmeijer E (2017) Heritage as sector, factor and vector: conceptualizing the shifting relationship between heritage management and spatial planning. Eur Plan
Stud 25(9):1654–1672

Kolen J (1995) Recreating (in) nature, visiting history. Second thought on landscape reserves and
their role in the preservation and experience of the historic environment. Archaeol. Dialog. 2
(2):127–159

Maggi M (2006) Ecomuseums worldwide: converging routes among similar obstacles. Chin Mus 3
(3):31–33

van Mensch P (2005) Nieuwe museologie. Identiteit of erfgoed? [New Museology. Identity of
heritage?]'. In: van der Laarse R (ed) Bezeten van vroeger. Erfgoed, identiteit en musealisering.
Het Spinhuis, Amsterdam, pp 176–192

Neal C (2015) Heritage and participation. In: Waterton E et al (eds) The Palgrave handbook of
contemporary heritage research. Palgrave Macmillan, Houndmills, pp 346–365

Opmeer M, Burgers GJ, Bruinsma R, Janssen R, Napolitano C, Ricci I (2019) Geospatial technologies in support of community enhancement and creating inclusive historical narratives. In:
Pyles D, Rish RM, Warner J (eds) Negotiating place and space through digital literacies :
research and practice. Information Age Publishing, Charlotte

Smith L (2006) Uses of heritage. Routledge, New York

Chapter 6
Analyzing Tourists' Preferences for a Restored City Waterway

Ila Maltese and Luca Zamparini

Abstract Up to 1929, Milan was labelled "Little Venice", due to its dense network of canals. In that year, the network was partially covered and replaced by roads aiming at faster freight and passenger transport. During an informal referendum held in 2011, most of Milan's citizens (95%) declared to be in favour of restoring the Navigli waterways system, now flowing underground in the centre of the city. Following this result, in 2015 the Polytechnic of Milan developed the feasibility study of the project. The waterways system could be reactivated for about 8 km, passing through the Eastern area of the city centre and connecting the North of the city to the South. In addition to the blueprint restoration, according to the sustainable mobility strategy adopted by the Municipality, a cycle-path will be running aside the canal, thus relieving congestion in high traffic and over-crowded roads. The Navigli restoration project would thus dramatically change not only the aesthetic character of the whole city of Milan but also the accessibility level—especially in the city centre—thanks to the new slow mobility infrastructure. Furthermore, this urban transformation would represent a big opportunity for the preservation and promotion of the cultural and historical heritage of the city, by partially restoring the ancient landscape. Ultimately, the city image would certainly benefit from the expected improvement of the overall quality of life generated by the project, especially concerning the "leisure" aspects. In order to explore the increased urban attractiveness, the value attached by tourists to the project has been investigated in the chapter. Both the cultural and environmental features of this urban transformation, together with the "non-users" benefits stemming from the slow mobility improvement, suggested to adopt a Total Economic Value (TEV) approach and a Contingent Valuation Method (CVM). To this aim, a survey was carried out in 2019 among 1070 tourists and excursionists in order to elicit their

I. Maltese
TRElab (Transport REsearch lab), Università degli Studi Roma3, Rome, Italy
e-mail: ila.maltese@uniroma3.it

L. Zamparini (✉)
Dipartimento di Scienze Giuridiche, Università del Salento, Lecce, Italy
e-mail: luca.zamparini@unisalento.it

preferences for the restored waterways system. The results have shown that visitors would be eager to spend a sensible amount of time to enjoy the waterfront and that they would mainly go there on foot or by bike, thus proving that the project would increase active mobility options and the consequent enhancement of transport sustainability in the city.

Keywords Urban attractiveness · Active mobility · Tourists' preferences · Waterways restoration · Total Economic Value · Contingent Valuation · Milan · Urban project

6.1 Introduction

Despite its historical business orientation, the tourism in Milan has recently become more based on leisure activities, particularly after the success of EXPO 2015, which attracted visitors from all over the world. Both arrivals and overnight stays have constantly risen until 2019, for all the considered origins of the visitors (Istat 2020).

Among the amenities of Milan (such as the Duomo or the Sforzesco Castle, to mention two of the most renowned ones), a leisure attraction, that appears to attract ever higher numbers of residents and tourists, is represented by the major canals (Naviglio Grande and Naviglio Pavese; hereinafter Navigli), departing from the newly reopened Darsena towards the countryside beyond the South of the city and running through very lively and attractive banks full of restaurants and cafes.

In the 1930s decade, an extended network of canals was covered in order to make freight and passenger transport faster and more comfortable, through high-speed roads. In 2011, the citizens of Milan were asked about the possibility to restore the Navigli canals network, by means of a referendum organized by the municipality. The "Yes" party gained 94.32% out of 450,000 votes. A clear indication of the need for a more sustainable mobility, for an improved urban landscape and quality of life. In the aftermath, a study group belonging to the Polytechnic of Milan has proposed a restoration project, which included a computation of the economic costs and an estimation, through Input-Output analysis and Hedonic Prices Methods (HPM), of the benefits, in terms of employment, economic growth, and real estate prices increase (Boscacci et al. 2015, 2017), despite the assumption that the project is not supposed to generate relevant direct revenues. Once realized, the restoration would also provide the city with a range of further benefits, including additional recreational activities; aesthetic improvement of the landscape thanks to the waterview (Cengiz 2013); slow mobility enhancement, with the consequent reduction of mobility-related negative externalities (congestion, air, and noise pollution) (Garrod and Willis 1994; O'Gorman et al. 2010). Moreover, such a cultural and environmental urban transformation would improve the general image of the city and it would increase its tourist attractiveness (Harrison and Sutton 2003).

In order to estimate these benefits, that would affect the current, future, and potential users as well as the non-users, a survey was administered, among residents and tourists. It aimed at assessing the value of the project, in terms of willingness to pay (WTP) for it, through the application of the Contingent Valuation Method (CVM).

Within this framework, the present chapter focuses on the perceived value of the project for tourists and excursionists, by analysing their answers to a structured questionnaire. In particular, it estimates their WTP by considering the stated hours that they would spend visiting the project while in Milan. To the best of the authors' knowledge, this is the first work to compute the value attached to a waterway project by means of time calculation stated preferences.

The chapter is structured as follows. After this introduction, few similar case studies on waterways restoration are presented. Data collection and methodology are then described, together with a brief project description, followed by the descriptive statistics results, while the last section provides some conclusions and policy recommendations.

6.2 Literature Review

This section provides a review of the literature by discussing some previous analyses of waterways projects and restorations. First, it must be noticed that most of the literature comes from project reports and valuation studies. Secondly, many cases of creation or restoration of previous existing canals network are well described from a qualitative point of view,[1] while they lack a thorough quantitative estimation of the benefits that may stem from these projects. Specifically, as concerns the evaluation methods, some studies are based on the consumers' revealed preferences for properties with a waterview, showing that they are more expensive,[2] while only few works focus on the direct assessment of the benefits of a new project in terms of the WTP for having it realized.

One of the first studies that has tried to analyse the value of the existing canal networks was related to Greater London and to the English Midlands, mainly privileging the point of view of the residents. It was carried out by means of an HPM together with a CVM survey administered among experts (Willis and Garrod 1993; Garrod and Willis 1994). Results showed that properties with a direct water frontage were valued 19% more than distant ones, while the premium of properties in a waterside development was 8%. Within this context, an assessment of the 26 km Bedford and Milton Keynes Waterways project in UK has been presented in 2009, focusing on the economic benefits of creating water frontage, including an increase in tourists' flows and consequent local expenditures (SQW Consulting 2009).

---

[1]Several cases of urban waterfront regeneration are described, among others, by Desfor et al. (2010), Timur (2013) and Hooimeijer (2014).

[2]See, for example, Chin and Chau (2003).

O'Gorman et al. (2010) have published a report on Inland Waterways, trying to fill the gap in the economic estimation of their intangible benefits mainly using the Benefit Transfer analysis. Moreover, they have proposed an extensive literature review on the categorization of benefits for the economic assessment of waterways. Similarly, Morris and Camino (2011) have used the same methodology for estimating benefits of inland and coastal wetlands. Moreover, Ouellette et al. (2014) analysed similar projects and interviewed several stakeholders in order to carry out a Cost Benefit Analysis of the restoration or replication of the Blackstone Canal in Worcester.

Buckman (2016) has proposed the Canal Oriented Development as a placemaking mechanism for helping cities in exploiting the benefits from their waterfront, while Masnavi et al. (2016) have applied a Fuzzy Analytical Hierarchy Process (FAHP) for the landscape river valley analysis in the Northern part of Tehran.

As concerns studies more focused on the impact of waterways on tourism, Mathis et al. (2004) have focused on the contribution to the sector of ecotourism in Texas Lower Rio Grande Valley in terms of economic value of the water (and not only its impact). They have also highlighted the difficulties in assessing the non-use values attached to the presence of the water from an aesthetic point of view. To this aim, the results of a survey incorporating CVM and Travel Cost Method (TCM) have been presented aiming at providing information about the ecotourists' (the consumers') WTP. Similarly, Savage et al. (2004) have examined the use of thematic zones, and specifically the river, in Singapore as a sustainable tourism strategy. More recently, a survey carried out among 702 visitors (local resident and non-resident) to the River Walk in San Antonio (Texas) between 2012 and 2013 (Nivin 2014) aimed to estimate the impact of the attractiveness of the waterways on the surroundings by means of an Input-Output model, jointly with a previous study related to the San Pedro Creek, in order to provide a wider network perspective (Nivin 2013). In Europe, Bernat has compared two different sized Polish cities in order to explore the relationship between the revitalization of their water sites and the development of tourism and recreation (Bernat 2014).

Given that the related bike lanes, the pedestrian corridors and the improved landscape that originate from a waterway project are also attractive for leisure time and tourism, some surveys have been carried out on the project value taking into account these connected infrastructures. In some cases, they have used a CVM, regardless of the presence of water canals. A case study, that is comparable to the project that constitutes the topic of this chapter, is related to the reopening of the Cheonggyecheon canal in Seoul by replacing an existing high-speed road. The analysis accounts for the expected increases of business activity (+3.5%) and of employment (+0.8%). However, the benefits emerging from the tourists' flows that would originate from the project have been only briefly mentioned. On the other hand, the sustainable mobility improvement in the city of Seoul from the project, due to a better pedestrian experience and attitudes, the increased liveability of the surroundings, the reduction of urban speed and, more broadly speaking, an improved quality of life have been taken into account by several studies related to this project (Klee 2006; Kang and Cervero 2009; Lee and Anderson 2013).

6.3 The Waterway Restoration Project in Milan

The restoration of Milan's old waterways system will connect Naviglio Martesana (in particular by Cassina de'Pomm), in the north of the city, with the ancient Darsena harbour and the two major canals (Grande and Pavese) in the South, covering an 8 km span in the centre of the city (Fig. 6.1) and including 25 bridges.

Furthermore, a slow mobility infrastructure (bike and pedestrian lane) is also planned to run along the restored canals. Together with a fast track lane for Local Public transport (LPT), it should relieve congestion in the current dangerous and polluted high-speed roads.

Apart from tackling mobility issues, the project also presents many cultural and historical aspects, which make it very interesting and attractive as a new amenity of the city.

Firstly, it must be mentioned that the Navigli network system, in particular with respect to its section flowing in the city centre, was reorganized in the first years of the sixteenth century by Leonardo da Vinci. Secondly, along the currently covered Navigli to be restored, there are many ancient buildings and other amenities that could be very attractive for the tourists.

The economic cost of the whole project was originally 406.6 million Euros, according to independent estimates (Boscacci et al. 2015). Further developments including a reconsideration of the VAT provided a more conservative estimate of 336.7 million Euros (Associazione Riaprire i Navigli 2018). It is estimated that the positive effects of this new canal would span for 500 m of distance from the canal bed. Moreover, expected outcomes of the project are represented by the possibility to increase freight and passenger transport options, to provide new leisure and recreation (on- and off-water) activities, and by to increase tourist flows and activities (such as cruises, new thematic museums, and art performances). It must be taken into account that specific theme tours are already proposed for the existing canals.

6.4 Methodology

As it was already mentioned, environmental and cultural benefits are expected to originate from the Navigli project. Even if the project does not exist yet, as a waterway restoration it concerns intrinsic, intangible, and non-monetary aspects such as cultural meaning, sense of urban identity (Hanh 2006; Piper 1997; Erham and Hamzah 2014), and landscape perception (VALUE 2012). Considering that these are all non–market goods, it could result very hard to estimate the overall value of the project (Pearce and Turner 1990).

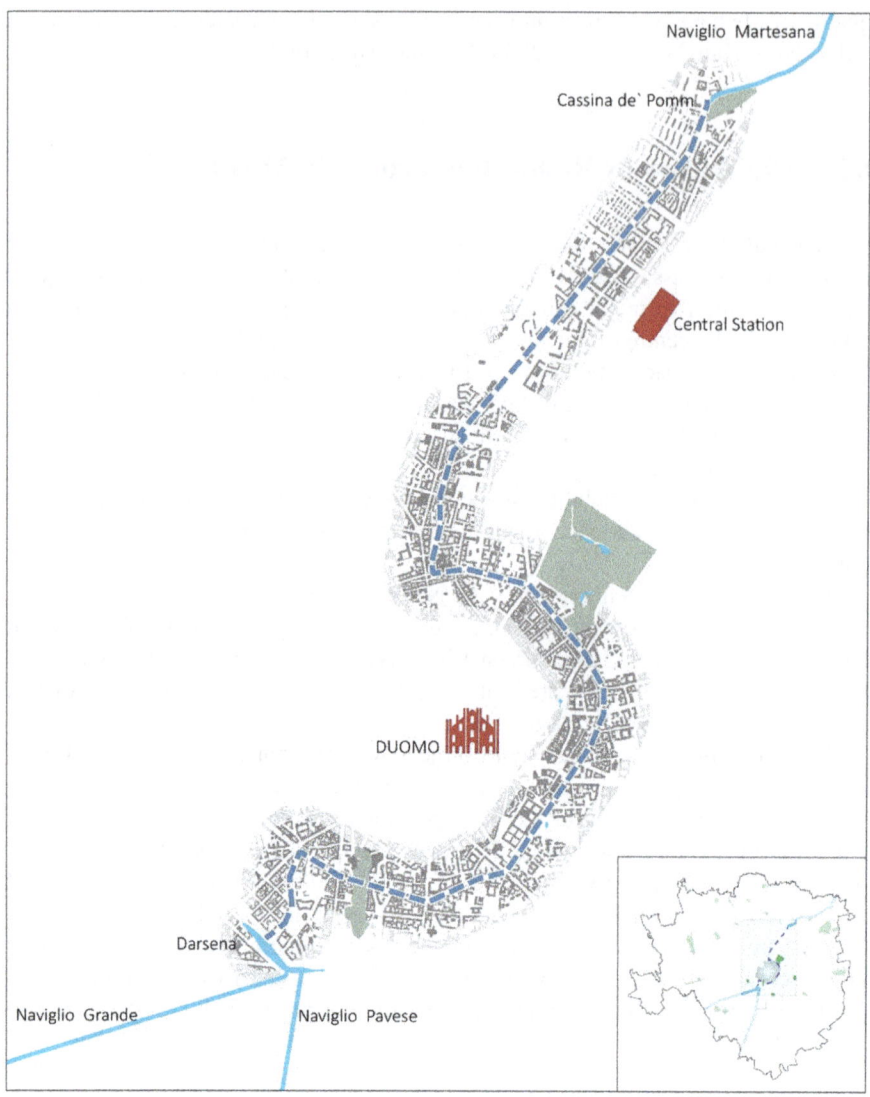

Fig. 6.1 The project blueprint. (Source: Boscacci et al. 2017)

Within this context, the use of Contingent Valuation Method (CVM) is strongly encouraged for two main reasons. Its well-acknowledged reliability (Arrow et al. 1993) and its capability to measure benefits from a non-market product, whenever non-use values (existence and bequest) are important value components (Mourato and Mazzanti 2002; Litman 2016).

More specifically, the CVM is a survey-based stated preference method, which evaluates a change from the *status quo* to a hypothetical ("contingent") market represented by the project to be assessed. Respondents are then required to assess how much they would be willing to pay for the project to be realized, i.e. the value they place on the good (Hausman 1993). This allows to estimate and trace a "latent" demand curve (bid curve) which may consider some explanatory variables (e.g. age, sex, income, education) collected during the survey. Some previous CVM studies already proved to be effective in assessing the intangible benefits coming from slow-mobility infrastructures (Maltese et al. 2017; Krizek 2007; Ruiz and Bernabé 2014; Vandermeulen et al. 2011; Lindsey and Knaap 1999). The conjoint use of CVM and Travel Cost Method (Betz et al. 2003) has proved effective in estimating visitors' demand on the basis of consumer surplus' measure. In particular, Viaud-mouclier (2012) has used CVM for assessing a cycling and walking path along the river Vesdre in Verviers (Province of Liege, Belgium).

In addition to the CVM, the Total Economic Value (TEV) approach has been also applied for capturing use (direct and recreational), potential use (option/safeguard), and non-use values (existence and bequest) of natural and cultural resources (Pearce and Turner 1990; Mourato and Mazzanti 2002) due to the concurrent existence of several values, for both current and future generations. When dealing with slow-mobility infrastructures like functional or recreational cycling paths, different categories of values have to be considered: use, potential use (in the future, maybe), and non-use (existence and bequest) (Litman 2016).

Lastly, there are many possible options (Ahmed and Gotoh 2006; Lopez-Feldman 2012) of questions aiming at estimating the WTP.

In the present chapter, where it will be quantified in hours instead of money, the CVM is based on an open-ended question. Each respondent was asked to state the specific amount of time he/she would be willing to spend for visiting the project site, once it will be realized.

6.5 The Survey

In order to measure the increase in attractiveness coming from the completion of the project, a survey among tourists and excursionists has been carried out in 2019 by means of a questionnaire exploring visitors' satisfaction about their stay in Milan, their potential visit to the uncovered Navigli, their knowledge, interest, and the value attachment to the project, and socio-demographic data.

In more detail, the test was composed of:

1. Section 1, where information about visitors' stay in Milan and the consequent level of satisfaction was gathered;
2. Section 2, where information about the visit to the existing Navigli and the consequent level of satisfaction was collected;
3. Section 3 aiming at investigating whether the interviewees knew the project—if not, a brief description and some pictures were provided—and whether they were in favour of it.

Due to the cultural and environmental nature of the project, a CVM was specifically adopted, in addition to the Total Economic Value (TEV) approach. Focusing on the CVM bidding game, an open-ended question has been chosen, which spared the pre-test stage.

Specifically, the visitors' willingness to pay for the project was expressed in the form of the additional time that they would be willing to spend in Milan for visiting the restored blueprint. On the other hand, the rating of the Total Economic Value (TEV) components (use, option, existence, and bequest values) has been requested.

4. Section 4, gathering the socio-demographic data of the respondents.

1088 face-to-face interviews were administered by well-trained students, who randomly selected a sample of visitors (both tourists and excursionists) in different urban locations of Milan. In most cases, the questionnaires were filled by the interviewers who took note of the answers that they received. In some cases (about 7%), the questionnaires were handed to those who agreed to participate in the survey while visiting Milan but preferred to complete the questionnaire by themselves. After some operations of data cleaning (−2%) which excluded some tests because the respondents were not a proper tourist or excursionist (e.g. off-campus student); therefore 1070 valid questionnaires (98.34%) were finally collected from the survey.

6.6   The Sample

The sample was composed both by Italian (88%) and foreign (12%) citizens. Fifty-four percent of respondents were male and only 26% live in a small family, made up of less than 3 people. Young people (i.e. with an age up to 30 years) are 48%, while only 7% are over 60. The other respondents are quite homogeneously distributed among the remaining age brackets 30–39, 40–49, 50–59.

Ninety-one percent of the sample holds at least a high-school degree, 39% a bachelor one, while 54% of the respondents are employed, in full- or part-time positions (Fig. 6.2) while the others are either unemployed or students or housewives

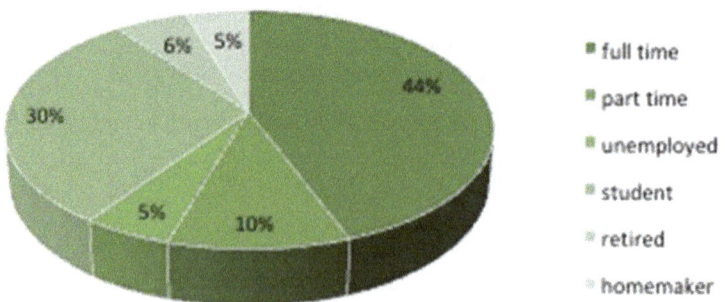

Fig. 6.2 Employment status of the sample

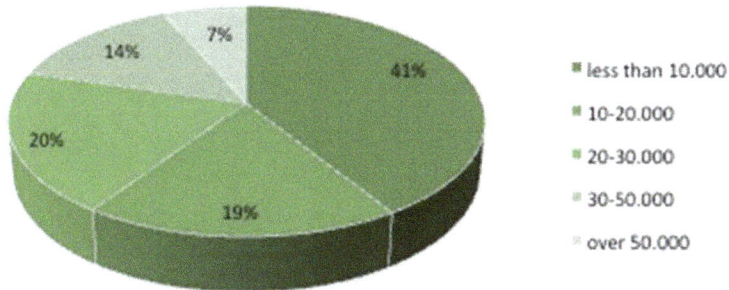

Fig. 6.3 Income of the sample

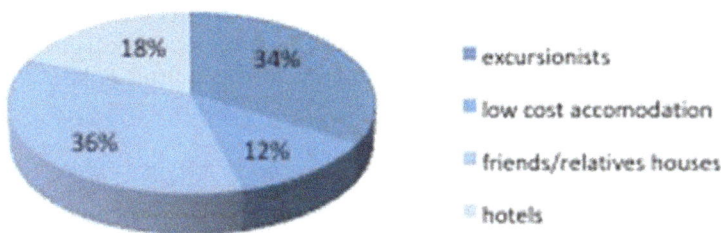

Fig. 6.4 Accommodation of the sample

or retired people. This is coherent with the stated yearly income that for 60% of the respondents is less than 20.000€ in a year (Fig. 6.3).

6.7 Descriptive Statistics

6.7.1 Visit to Milan

As concerns the trip to Milan, the city was reached by train (48%), by plane (24%), by car (24%), and by bus (4%). The trip purpose was only leisure for 72% of the sample, business, study, or health reasons for 9%, and a mix of personal and leisure motives for the remaining 19%.

When the travelling group is considered, 32% of the respondents were travelling to Milan alone, 22% were part of an organized trip, another 22% were with a group of colleagues, 18% were travelling with a partner, and 6% with family or friends.

Sixty-six percent of the sample stayed in Milan at least for one night; in particular 36% was hosted by friends or relatives or used a second home, while the other 30% was in a paid accommodation in the city (Fig. 6.4).

Excluding the excursionists, the average stay in Milan is 2.66 days, with the modal value of two nights, while the average daily expenditure for a day in Milan for

Table 6.1 Visit at the Navigli

Day period		Week period		Season	
Day	22%	Weekday	22%	Spring/Fall	21%
Night	47%	Holiday	49%	Summer	13%
Both	30%	Both	29%	Winter	24%
				All seasons	42%

Source: Authors' elaboration

the whole sample[3] is about 71€. Finally, 51% of the interviewees think that Milan is a liveable city. For 22% of the visitors it was the first time in Milan, while 49% have been in the city at least six times.

Moreover, also due to the limited size of the city, and the consequent density of the amenities, the respondents have been, at least outside or in the neighbours of eight places, on average. The most visited site is the Duomo (98%), more than 80% visited the Sforza's Castle, 77% has been at least in front of the Scala theatre. More modern tourist spots, as the Fashion streets or the Gae Aulenti square, have been visited by 70% of the sample.

6.7.2   Visit to Navigli

As concerns specifically the knowledge of the Navigli, the two canals in the south (Naviglio Grande and Pavese) and their connecting ancient harbour (the Darsena), which are a hotspot for the nightlife, are well known by almost 90% of the respondents. Conversely, the Naviglio Martesana in the north, quieter and with a bike lane on its side, is known by 54% of the sample, while only 33% has been there. Twenty-eight percent of the sample has never been on any Naviglio canal; among the left 72%, 43% are repeaters, i.e. have been there more than once, 29% have been only once on the canals. In more detail, 47% reached the canals only during the night and 22% only during the day; 49% in the weekend, 22% not in the weekend; 24% only in the winter, 13% only in the summer; 21% only in fall or spring (Table 6.1).

6.7.3   The Restoration Project

A last set of questions concerned the project for the restoration. First of all, it must be noticed that only 23% of the non-residents knew the project.

Once provided with a brief description and some pictures/rendering, the sample believes that the restoration project will benefit both residents and tourists (61%), only tourists (15%), or residents (7%). Only 2% of the sample cannot see any benefit, due to the fear of increased congestion in the neighbour roads or to the cost

---

[3]The question was about the expenditure in a day not including travel or staying cost, thus being equivalent for tourist and for excursionist.

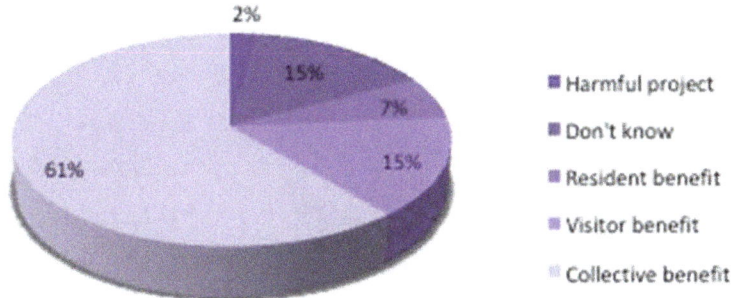

Fig. 6.5 Benefits from the project

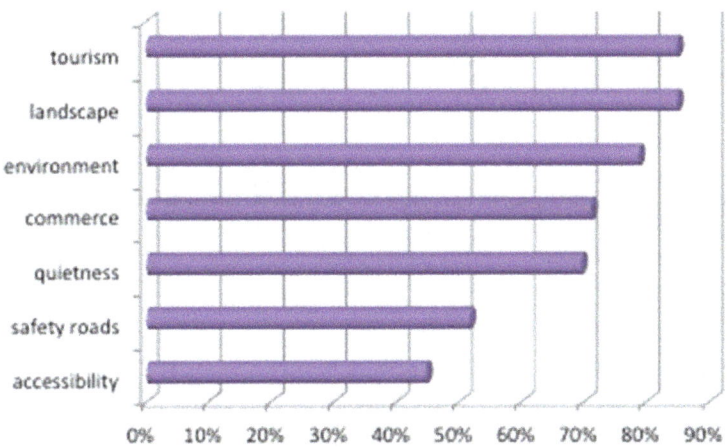

Fig. 6.6 Impact of the project

opportunity of the public investment (i.e. the need to put the public resources in different projects/investments)[4] (Fig. 6.5).

In particular, the respondents identified on average five positive impacts of the project, mostly on landscape (85%) and tourism (85%), environment (79%), while accessibility is positively impacted for less than the half of the sample (45%) (Fig. 6.6).

Asked about their potential use of the infrastructure, apart from 27% of waverers/hesitants and 2% of not interested, more than half of the sample would only walk along the canals (53%), while 13% could enjoy those places both by foot and by bike. The left 6% would only cycle throughout the renewed landscape of the city (Fig. 6.7).

---

[4]As a reply to these doubts many scholars are being working on traffic modelling and cost benefit analysis; see for ex. Goggi and Indelicato (2015) or the material available on the official website of the project (Municipality of Milan) https://progettonavigli.comune.milano.it/materiali/analisi-costi-e-benefici/

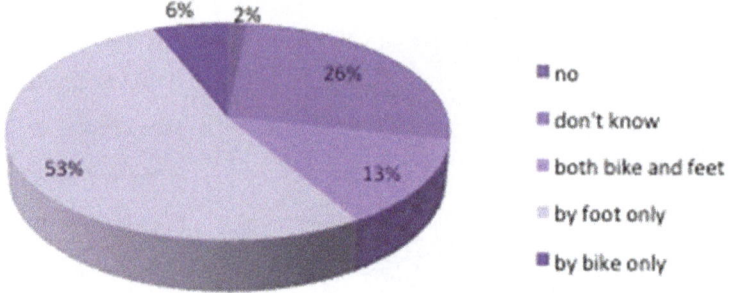

Fig. 6.7 Possible use of the project

Table 6.2 Likert scale for the TEV components

	Use (%)	Existence (%)	Bequest (%)	Safeguard (%)
1	2	3	3	1
2	8	11	8	5
3	32	32	26	23
4	33	33	32	35
5	25	22	31	36

Source: Authors' elaboration

6.7.4   Total Economic Value

The importance given to specific aspects related to the project has been expressed by the respondents through a five-point Likert scale, where 1 expresses no importance and 5 represents the highest level of relevance. As illustrated in Table 6.2, for all the elements of the TEV, those attaching a low significance (i.e. a value of 1 or 2) are only a scarce minority.

Safeguard (i.e. the preservation of the environment and landscape by means of the restoration) is considered as the most relevant dimension of the restoration project. It is deemed important or very important by 71% of the respondents. Sixty-three percent of the visitors believe that it is at least important to leave such a restoration project to the future generation (Value of bequest). The possibility to make use of the infrastructure and to directly take advantage from the restored canals network is important or very important for 58% of the respondents. Lastly, 55% states a value of 4 or 5 for the mere existence, i.e. the awareness and knowledge of the canals' existence, even without using them.

6.7.5   Contingent Valuation Method

According to the CVM Guidelines (Arrow et al. 1993), when posing an open-ended question, it can be difficult for the respondents to identify their own WTP, especially when the good or project does not exist yet. Furthermore, there is also the possibility that they behave strategically or reply for pleasing the researcher. In order to minimize these possibilities, the respondents to the survey were asked to state the amount of time that they would be willing to spend visiting the restored Navigli areas. Moreover, given that a similar place already exists, and it is quite well known in the city of Milan, a clear description and explanation of the project was provided at the beginning of the test administration.

The results show that on average the tourists and excursionists would spend 4 h more for visiting the restored canals (2 h is the modal value, while the median value is 3), enjoying the new landscape or making use of the slow mobility infrastructures.

The minimum value of 0, differently from the case of WTP expressed in terms of money, cannot be considered as a proxy of the lack of interest for the project. Indeed, among those responding that they would not be going to spend their time on the Navigli restored canals (less than 3%), 80% are not opposing the project, considering its positive impact (55%) and attaching at least medium values to the TEV components (70%).

Plausibly, only 2% of the sample would spend more than half a day on the project; out of these 20 respondents, most (i.e. 65%) would stay 1 day more in Milan for visiting the new Navigli blueprint, 20% (4) declared that they would prolong their stay by 2 days, 5% (1) by 3 days, while the two left would stay 4 and 6 days more; indeed 150 h was the maximum amount of hours in the sample.

That said, it has also been possible to trace the bid curve (Fig. 6.8), where the x-axis illustrates the visitors' number, while y-axis represents the amount of time they would spend visiting the project, expressed in hours.

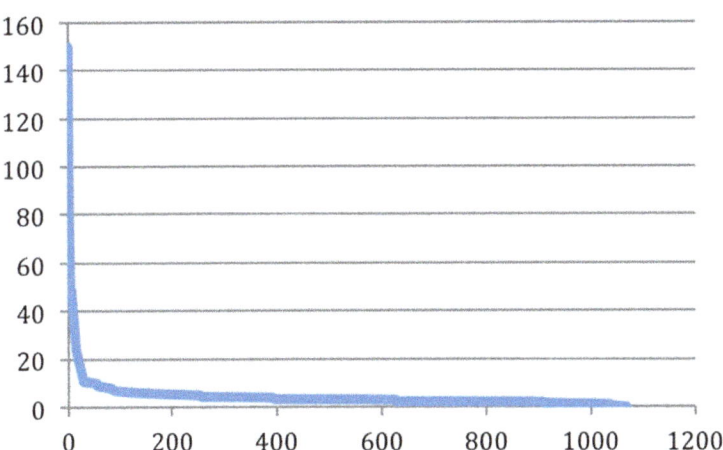

Fig. 6.8 Bid curve. (Source: Authors' elaboration)

6.8 Conclusions

In this chapter, the tourists' bid curve expected from the project of the Navigli restoration in Milan has been estimated and traced through the CVM, by directly asking tourists to state their WTP—in terms of hours possibly spent for visiting the restored Navigli network—for the project. Moreover, they have been also investigated about the potential use (option/safeguard value) and non-use (existence and bequest values) components of the TEV. Both the analysis of the WTP in hours and the direct investigation of the single TEV components show a relevant approval rate among the visitors of Milan. It would be interesting to consider also these results—together with those provided by the test administration among the residents—when considering the benefits that could be generated by the project.

From a methodological point of view, this "time-based" application proved to reduce the potential CVM limits and biases. More specifically, not only most of the respondents already knew or have been on the existing Navigli, but also they were provided with pictures and detailed description, thus minimizing the "information bias". Furthermore, spending time instead of money seems easier to state, avoiding the possible misuse of the test as a means to give voice to protest and objection against the project. At the same time, the strategic bias which can push toward the realization of the project has also been minimized by the control questions about the TEV and the wish to take advantage of the project, once it will be realized.

These results justify the willingness to accomplish the project, which was also strongly endorsed by residents, as witnessed by the results of the referendum among the citizenship. Furthermore, it could represent a good strategy for achieving a more sustainable mobility in the city, based on active travel—which in the aftermath of Covid-19 pandemic seems to be preferred also to LPT (Maltese et al. 2021)—and on slow(er) and safe(r) private motorized traffic, at the same time.

Lastly, it must be considered that the project constitutes an attraction for tourists and excursionists but also an infrastructure for residents. These different stakeholders can jointly use the slow mobility infrastructure by foot or by bike or simply enjoy the improved landscape and environment. This common space/infrastructure may thus give rise to the possibility for tourists/excursionists to meet and share activities and experiences with residents. This may turn out as a relevant marketing opportunity for the city of Milan making the city, usually perceived as a business tourist destination, more attractive also for leisure purposes.

Acknowledgments The authors would sincerely like to thank all the respondents, for giving their time and information and the students of the Course of Environmental Economics and Sustainability at the Polytechnic of Milan, who carried out the in-person interviews.

References

Ahmed US, Gotoh K (2006) Cost-benefit analysis of environmental goods by applying contingent valuation method. Springer, Tokyo

Arrow K, Solow R, Portney PR, Leamer EE, Radner R, Shuman H (1993) Report of the NOOA Panel on Contingent Valuation. Fed Regist 58:4601–4614

Associazione Riaprire i Navigli (2018) Riaprire i Navigli a Milano. Le modalità di finanziamento del progetto e le ricadute socioeconomiche dell'investimento. Fondazione Cariplo, Milan. https://www.riaprireinavigli.it/uploads/Report_Navigli_WEB_.pdf

Bernat S (2014) Revitalization of river valleys and development of tourism (Lublin and Puławy Study Case). Geogr Tour 2:7–16

Betz CJ, Bergstrom JC, Bowker JM (2003) A contingent trip model for estimating rail-trail demand. J Environ Plan Manag 46(1):79–96

Boscacci F, Camagni R, Caragliu A, Maltese I, Mariotti I (2015) Valutazione dei benefici collettivi. Considerazioni finali sui costi e sui benefici dell'intervento. In: Boatti A. et al. (Ed.) Attività di ricerca scientifica e Tecnica finalizzate allo Studio di fattibilità per la riapertura dei Navigli Milanesi nell'ambito della riattivazione del sistema complessivo dei Navigli e della sua navigabilità, pp 125–154. Available via RE.PUBLIC@polimi (Research Publications at Politecnico di Milano) https://re.public.polimi.it/handle/11311/986050#.X-iwJelKi01. Accessed 27 Dec 2020

Boscacci F, Camagni R, Caragliu A, Maltese I, Mariotti I (2017) Collective benefits of an urban transformation: restoring the Navigli in Milan. Cities 71:11–18

Buckman S (2016) Canal oriented development as waterfront place-making: an analysis of the built form. J Urban Des 21(6):785–801

Cengiz B (2013) Urban river landscapes. In: Advances in landscape architecture. IntechOpen, London

Chin TL, Chau KW (2003) A critical review of the literature on the hedonic price model. Int J Hous Appl 27(2):145–165

Desfor G, Laidley J, Stevens Q, Schubert D (eds) (2010) Transforming urban waterfronts: fixity and flow. Routledge, London

Erham A, Hamzah A (2014) The morphology of urban waterfront tourism: the local identity portray in Melaka and Makassar. Urban Econ Reg Stud eJ 7:155

Garrod G, Willis K (1994) An economic estimate of the effect of a waterside location on property values. Environ Resour Econ 4:209–217

Goggi G, Indelicato V (2015) 5.1 La riapertura dei Navigli nel sistema della mobilità Milanese. In: Boatti A. et al. (Ed.), Attività di ricerca scientifica e Tecnica finalizzate allo Studio di fattibilità per la riapertura dei Navigli Milanesi nell'ambito della riattivazione del sistema complessivo dei Navigli e della sua navigabilità, pp 125–154. Available via RE.PUBLIC@polimi (Research Publications at Politecnico di Milano) https://re.public.polimi.it/handle/11311/986050#.X-iwJelKi01. Accessed 27 Dec 2020

Hanh VTH (2006) Canal-side highway in Ho Chi Minh City (HCMC), Vietnam–issues of urban cultural conservation and tourism development. GeoJournal 66(3):165–186

Harrison AJM, Sutton RD (2003, March) Why restore inland waterways? In: Proceedings of the Institution of Civil Engineers-municipal engineer, vol 156, no 1, pp 25–33. Thomas Telford Ltd

Hausman JA (ed) (1993) Contingent valuation: a critical assessment. North Holland Press, Amsterdam

Hooimeijer F (2014) More urban water: design and management of Dutch water cities. CRC Press, Cleveland

Istat (2020) Arrivals and night overstays. i.Stat. Accessed 27 Dec 2020

Kang CD, Cervero R (2009) From elevated freeway to urban greenway: land value impacts of the CGC project in Seoul, Korea. Urban Stud 46(13):2771–2794

Klee IK (2006) Cheong Gye Cheon restoration project - a revolution in Seoul. ICLEI conference proceedings, Capetown, 27 Feb–3 March 2006. http://worldcongress2006.iclei.org/uploads/media/K_LEEInKeun_Seoul_-_River_Project.pdf. Accessed 27 Dec 2020

Krizek KJ (2007) Estimating the economic benefit of bicycling and bicycle facilites: an interpretive review and proposed methods. In Essays on transport economics, pp 219–248. https://doi.org/10.1007/978-3-7908-1765-2-14

Lee JY, Anderson CD (2013) The restored Cheonggyecheon and the quality of life in Seoul. J Urban Technol 20(4):3–22

Lindsey G, Knaap G (1999) Willingness to pay for urban greenway projects. J Am Plann Assoc 65(3):297–313

Litman T (2016) Evaluating active transport benefits and costs. In: Guide to valuing walking and cycling improvements and encouragement programs. Victoria Transport Policy Institute, 26 March 2016

Lopez-Feldman A (2012) Introduction to contingent valuation using Stata. Munich personal RePEc archive: Centro de Investigacion y Docencia Economicas (Cide), MPRA WP 41018

Maltese I, Gatta V, Marcucci E (2021) Active Travel in Sustainable Urban Mobility Plans. An Italian overview. Res Transp Bus Manag 100621

Maltese I, Mariotti I, Oppio A, Boscacci F (2017) Assessing the benefits of slow mobility connecting a cultural heritage. J Cult Herit 26:153–159

Masnavi MR, Tasa H, Ghobadi M, Farzad Behtash MR, Negin TS (2016) Restoration and reclamation of the river valleys' landscape structure for urban sustainability using FAHP process, the case of northern Tehran-Iran. Int J Environ Res 10(1):193–202

Mathis M, Matisoff D, Pritchett T (2004) The economic value of water for ecosystem preservation: ecotourism in the Texas lower Rio Grande Valley. Final report to the Texas Coastal Management Program, General Land Office Contract 03-020

Morris J, Camino M (2011) Economic assessment of freshwater, wetland and floodplain (FWF) ecosystem services. UK National Ecosystem Assessment Working Paper. Cranfield University, Bedford

Mourato S, Mazzanti M (2002) Economic valuation of cultural heritage: evidence and prospects. In: De la Torre M (ed) Assessing the values of cultural heritage research report. The Getty Conservation Institute, Los Angeles, pp 51–76

Nivin S (2013, December) Economic and fiscal impacts of The San Pedro Creek improvements. https://fx8ez2bioiz3t7vbo9jbdd5l-wpengine.netdna-ssl.com/wp-content/uploads/2018/04/San_Pedro_Creek_economic_impact_report_-_final_2.pdf. Accessed 27 Dec 2020

Nivin S (2014, April) Impact of the San Antonio River walk. San Antonio Tourism. https://silo.tips/download/impact-of-the-san-antonio-river-walk Accessed 27 Dec 2020

O'Gorman S, Bann C, Caldwell V (2010) The benefits of inland waterways (2nd edn). A report to Defra and IWAC. Reference number, WY0101

Ouellette DJ, Messier JS, Spunar NS, Crimmins RC (2014) Cost and benefit analysis of Blackstone Canal revitalization. Worcester Polytechnic Institute, Worcester

Pearce D, Turner R (1990) Economics of natural resources and the environment. Harvester Wheatsheaf, London

Piper TR (1997) Central Canal walk, broad ripple, Indiana: urban waterfront reclamation: the creation of identity and sense of place through the revitalization of an urban waterfront. Cardinal Scholar. https://cardinalscholar.bsu.edu/handle/handle/188784. Accessed 27 Dec 2020

Ruiz T, Bernabé JC (2014) Measuring factors influencing valuation of non motorized improvement measures. Transp Res A 67:195–211. https://doi.org/10.1016/j.tra.2014.06.008

Savage VR, Huang S, Chang TC (2004) The Singapore River thematic zone: sustainable tourism in an urban context. Geogr J 170(3):212–225

SQW Consulting (2009) Bedford & Milton Keynes Waterway Economic Impact Assessment. www.sqw.co.uk. Accessed 27 Dec 2020

Timur UP (2013) Urban waterfront regenerations. In: Advances in landscape architecture. IntechOpen, London. Accessed 27 Dec 2020

Value (2012) The Value project – final report. South Yorkshire Forest Partnership/Sheffield City Council. http://www.value-landscapes.eu/. Accessed 7 March 2016

Vandermeulen V, Verspecht A, Vermeire B, Van Huylenbroeck G, Gellynck X (2011) The use of economic valuation to create public support for green infrastructure investments in urban areas. Landsc Urban Plan 103:198–206

Viaud-mouclier C. (2012). Valuing attractive landscapes in the urban economy - cycling and walking path along the river Vesdre. Final Report, March 2012

Willis KG, Garrod GD (1993) The value of waterside properties: estimating the impact of waterways and canals on property values through hedonic Price models and contingent valuation methods. Countryside change initiative working paper 44. Department of Agricultural Economics and Food Marketing, University of Newcastle upon Tyne

Chapter 7
Space Invaders? The Role of Airbnb in the Touristification of Urban Neighbourhoods

Bart Neuts, Karima Kourtit, and Peter Nijkamp

Abstract The rise of peer-to-peer booking networks for accommodations such as Airbnb has significantly altered the tourism landscape and has had broad implications for destination management. Apart from providing direct competition to the traditional hospitality sector, concerns have been raised about urban gentrification effects—through an influence on housing prices—and neighbourhood touristification. Subsequently, many European cities have started to design policies in order to control or limit Airbnb activities in their destination. This study investigates the impact of Airbnb in the city of Amsterdam on: (a) the average housing value per m^2 and (b) the business growth in a tourism-centric local economy. A large dataset is collected from Inside Airbnb on the supply of Airbnb listings and associated hosts. Neighbourhood data on demographics, as well as on housing, spatial, and economic indicators, was collected from the official statistical agencies of Amsterdam. The dataset spanned 5 years (2015–2019) and was collected at the detailed level of 99 neighbourhoods in the city. Fixed effects Panel-Corrected Standard Errors models were used and a significant effect of Airbnb listings was found on both the property value per m^2 and the touristification of neighbourhoods—modelled as the percentage of tourism-related activities in total neighbourhood business activities. At the same time, other explanatory variables appeared to be equally significant, and therefore contradict a myopic view on the drivers of neighbourhood touristification.

Keywords Airbnb · Peer-to-peer networks · Tourism-led gentrification · Touristification · Fixed effects

B. Neuts (✉)
Department Earth and Environmental Sciences, KU Leuven, Leuven, Belgium
e-mail: bart.neuts@kuleuven.be

K. Kourtit · P. Nijkamp
Open University, Heerlen, The Netherlands

Alexandru Ioan Cuza University, Iasi, Romania

Polytecnic University, Ben Guerir, Morocco

Adam Mickiewicz University, Poznan, Poland

© The Author(s), under exclusive licence to Springer Nature Singapore Pte Ltd. 2021   103
S. Suzuki et al. (eds.), *Tourism and Regional Science*, New Frontiers in Regional Science: Asian Perspectives 53, https://doi.org/10.1007/978-981-16-3623-3_7

7.1 Introduction

In the 'urban century' urban agglomerations and big city are increasing becoming economic, knowledge, technological, political and cultural 'magnets' in open and globalizing society (Glaeser et al. 2020). Tourism plays a central role in this megatrend. Large cities have a long history as tourist catchment areas, be it for leisure, business activities, cultural or sports events, visiting friends and relatives, or other related purposes. However, only in the last four decades have cities actively pursued tourism as an industry of importance. The increasingly globalized industrial system that has led to a relocation of many manufacturing firms from Western cities to lower wage countries has largely converted the economic base of urban areas towards finance, business services, tourism, and creative industries (Fainstein 2008). Apart from exerting regenerative effects on the region's economy (Hall and Page 2009), countering the loss of jobs and the increase in inner-city derelict spaces caused by deindustrialization, tourism was also considered a low-impact ('smoke-less') growth industry (Law 1992). As noted by McKercher (1993: 135–136) on the subject of sustainable tourism development: 'Some sectors of the industry are essentially immune to its effects. Existing large developments, virtually the entire spectrum of urban tourism and a wide range of other tourism activities will probably not be affected'. This quote seemingly implies that the urbanity of a destination offsets possible negative impacts incurred by tourist resource use, since cities are generally multifunctional and dynamic systems with robust infrastructures capable of dealing with large number of populations.

The view on urban tourism as an urban economic growth motor with limited direct negative impacts has directed entrepreneurial strategies in order to specifically improve the competitive advantage of the city as an attractive tourist destination (Daskalopoulou and Petrou 2009; Fusco Girard and Nijkamp 2009; Giaoutzi and Nijkamp 2017; Malecki 2002; Pavlic et al. 2013). While the rise in urban tourism has undoubtedly contributed to urban wealth creation (Neuts 2020), the increasingly competitive spatial claims of diverse user groups—in its most simplistic sense, residents versus visitors—have led to issues about long-term sustainability of such growth strategies. Studies by, for instance, Dodds (2007), du Cros (2007), Gilbert and Clark (1997), Gursoy and Rutherford (2004) and Pérez and Nadal (2005), are among the earlier examples to identify (potential) negative impacts of tourism in cities. The tourism impacts that are most significant in urban research largely concern the social dimension, seemingly implying a need to reconceptualize sustainability from its original 'resource-based' perspective—which is mainly grounded in the ecological dimension—to a 'norm-based' perspective—with a strong focus on the sociocultural dimension (Neuts 2016). Recent years, in particular, have seen a strong rise in negative sentiments surrounding urban tourism. Primarily—although not uniquely—in European destinations, the concept of 'overtourism' has been introduced to refer to a situation in which negative implications of tourism, borne by the local residents of a city, seemingly outweigh the positive socio-economic benefits (Koens et al. 2018).

One of the potential negative impacts of urban tourism growth relates to the commodification of space, where people and their local cultures and lifestyles are increasingly becoming tourism resources (Meethan 2001). Further driven by contemporary motivations for 'authentic' experiences and meaningful interactions (Füller and Michel 2014; Paulauskaite et al. 2017), tourism extends beyond a historical tourist bubble, further intertwining recreational and daily residential activities across neighbourhoods (Pappalepore et al. 2014).

This book chapter investigates property value increases and 'touristification' of primarily residential neighbourhoods in Amsterdam, through the use of a panel study. A particular interest lies in the potential contribution of online peer-to-peer platforms (in this study: AirBnB) as a main driver of rising housing prices—due to the investment potential offered by Airbnb listings—and an expansion of tourism beyond the traditional tourist core (Koens et al. 2018)—since these platforms are (a) more difficult to manage through urban planning, (b) play into the quest for authentic urban tourism experiences (Paulauskaite et al. 2017), and (c) are by their very nature linked to residential housing. The next section discusses the concepts of tourism precincts and touristification, followed by a section on Airbnb effects in tourism cities. Next, the methodology section introduces the data and statistical methods being used in the chapter, after which the results and concluding section interprets the empirical findings on temporal and spatial dimensions.

7.2 From Tourism Precincts to Tourism Cities

As noted by Page and Hall (2003: 20) it is difficult to univocally define urban tourism spaces since: 'tourists are attracted to cities because of the specialized functions they offer and the range of services provided'. Jansen-Verbeke (1986) identified primary, secondary, and conditional (or additional) elements within the urban tourism systems. Primary elements can be subdivided in activity places and the leisure setting, with the activity places referring to the supply side of the main tourist attractions such as concert halls, museums, exhibitions, sports facilities, casinos, and nightclubs. Regularly organized festivities and events are also considered elements of primary tourist activity places. The leisure setting, as the second dimension of primary elements, refers to the wider urban network of both physical, tangible heritage—such as monuments and statues, historical buildings, parks and green areas, waterfronts, etc.—and intangible features of local customs, atmosphere and ambience, way of life, language, and clothing. Compared to the activity places of tourism, the leisure setting is therefore not a unique and specific place of tourism consumption as such, but rather gives the urban destination its attractive overlaying image (Jansen-Verbeke 1986). Secondary elements refer to supporting facilities and services that, while being consumed by tourists, are generally not central to the travel motivations. Secondary elements of note are accommodation facilities and shopping facilities (Jansen-Verbeke 1986; Page and Hall 2003). However, Shaw and Williams (1994) also note how for many urban destinations, such secondary elements might actually be among the main tourist attractions, particularly for certain tourism segments, therefore blurring the distinction between primary and secondary tourism resources. Finally, conditional or additional elements influence the physical ability to travel, rather than the reasons for travelling. This includes, for instance, the state of accessibility, transportation and parking facilities, availability of tourist information, and signage (Jansen-Verbeke 1986).

Even though urban spaces—and particularly a city's public spaces—cater to varied groups of urban dwellers, resulting from either historical development (Ashworth and Tunbridge 2000; Jansen-Verbeke and Lievois 1999), policy choices (Maitland 2008), or a combination of both, leisure products tend to cluster in specific tourism precincts (Kelly 2008). Such zoning of tourist functionalities can increase attractiveness, accessibility, and utility for visitors by offering a multifunctional recreational space in a contained urban area (Hasegawa 2010; Jansen-Verbeke and Lievois 1999). Furthermore, even though from a visitor perspective, there might not be an overt demarcation of tourism versus non-tourism precincts, tracking studies in urban areas have shown how tourists generally limit themselves to specific, concentrated areas (e.g. Asakura and Iryo 2007; Keul and Kühberger 1997; Shoval 2008).

Further exacerbated by strong increases in visitor numbers throughout the last decades, potential conflicts have started to arise from the heterogeneous demands on urban space, with an increased competition for resources leading to road congestion, longer waiting times, increased prises, and crowding-out of residential functions (Law 2002). The tourism precinct has seemingly gradually transformed into a uni-functional area that insufficiently caters to local residents. Gravari-Barbas and Guinand (2017) define this process of tourism-based transformation of urban environments as 'touristification', which Freytag and Bauder (2018) identify according to three layers: a visual, non-visual and behavioural layer. The visible process of touristification entails changes in the built environment and/or functional use of the area, noticeable through the increase of tourism-specific services and stores. Non-visible—or hardly visible—changes pertain to modified use of existing infrastructure, such as the conversion of rental properties to short-term holiday rentals, either temporarily or continuously. Finally, the behavioural layer focuses on existing tourist practices and can further lead to perceptual changes in local residents on the area's primary functionality (i.e. transforming from a residential space to a recreational space).

Ojeda and Kieffer (2020) state how touristification—particularly from the viewpoint of tourism geography—ought to refer to the process by which socio-economic dynamics within a given territory change to give rise to a tourism-dominant local economy. They criticize the fraught conceptualization of the term, which is sometimes used synonymously with gentrification or as an intrinsically negative attitudinal state of the host population. Such use erroneously equates touristification with other concepts like 'carrying capacity', 'overtourism', or 'limits to change', as is clear in the definition adopted by Barrera-Fernández et al. (2019: 103): 'Touristification is the condition by which a city or another type of tourist destination such as a beach or a natural park receives a number of visitors that make the residents' quality of life and the quality of the tourist experience deteriorate in an unacceptable manner'.

Touristification processes are intrinsically linked to the existence of (historic) tourist precincts in a circular relation: a tourist precinct exists through the original clustering of tourist attractions and secondary support systems (i.e. already being 'touristified'), attracting additional tourism-related services to the area and leading to a further monopolization of the economic base (i.e. increasing 'touristification' of the area) (Sequera and Nofre 2018). However, recent years have also seen the rise in touristification of urban neighbourhoods that were long considered 'off the beaten track' and outside of main tourist-centric areas; examples include Kreuzberg, Neukölln, Prenzlauer Berg, and Friedrichshain in Berlin (Novy 2016, 2018), the Lombok neighbourhood in Utrecht (Ioannides et al. 2019), London's East End district (Maitland 2006; Maitland and Newman 2004; Shaw et al. 2004), the area around Canal Saint Martin in Paris (Gravari-Barbas and Jacquot 2016), and Gracia in Barcelona (Fava and Rubio 2017). These processes were at least partly supported by (a) policymaking—in an attempt to spread tourism across the city—, (b) contemporary tourism demands for local experiences and authentic lifestyles (Nilsson 2020; Paulauskaite et al. 2017), (c) social media—decentralizing destination marketing and planning—, and (d) the rise of peer-to-peer accommodation services such as Airbnb (Freytag and Bauder 2018).

7.3 The Role of Airbnb in Urban Tourism

In recent years, the rapid rise of peer-to-peer platforms in the field of tourism has significantly altered the landscape of urban tourism and tourism planning. While occasionally lauded as an instrument towards democratization of tourism receipts towards a larger share of the destination population (Kadi et al. 2019) and seen as an additional segment of the sector to strengthen the total economic effects of tourism, cities have increasingly raised concerns about the disruptive effects of a wide-spread use of such platforms.

Focusing specifically on Airbnb as the most successful peer-to-peer accommodation platform—and in many destinations achieving a near monopoly due to the strong network effects present—studies have focused primarily on the relationship between Airbnb and the traditional hotel sector (e.g. Choi et al. 2015; Heo et al. 2019; Zervas et al. 2017) or the spatial relationship between Airbnb and the destination place (Gutiérrez et al. 2017; Heo et al. 2019; Ioannides et al. 2019; Xu et al. 2020). Such studies have, however, not led to general agreement on the impacts of Airbnb. While Zervas et al. (2017) noted a causal impact of up to 8–10% on hotel revenues in Austin, Texas—primarily through price responses of traditional hospitality providers—Choi et al. (2015) concluded that there was no effect on hotel revenue in the Republic of Korea, also noting that Airbnb has a low awareness rate in this particular location. Linking the effect on hotel prices with the spatial distribution of Airbnb accommodation, Heo et al. (2019) suggest that traditional providers and peer-to-peer platforms are not in direct competition and that there are dissimilarities in geographic location and seasonality patterns, at least in their case study of Paris. On the other hand, Gutiérrez et al. (2017) studied the location patterns of Airbnb in Barcelona and found that Airbnb is dominant around the city's hotel axis while also remaining close to the city's main tourist attractions. These findings are shared by Xu et al. (2020) in the case of London, where Airbnb listings were mainly located in the city centre and around tourist attractions. Ioannides et al. (2019) investigated the neighbourhood impacts of Airbnb in Utrecht—a secondary tourist city in the Netherlands—and found some evidence of a tourist gentrification effect.

The latter cases are of particular interest, since there seems to be some conflicting evidence. On the one hand, Airbnb have themselves stated in corporate communication that they help to spread tourism away from congested urban areas (Airbnb 2019), while also supporting local people through economic empowerment (Business Insider 2016)—which can then potentially increase touristification and tourism-led gentrification in such non-central neighbourhoods. On the other hand, if—as found by Gutiérrez et al. (2017) and Xu et al. (2020)—Airbnb properties are found to cluster around established tourist areas, the result would rather be a further concentration of tourism, potentially contributing to overtourism at such locations. Therefore, the present book chapter attempts to further investigate the impacts of Airbnb listings from a spatial and temporal perspective by analysing the effects of Airbnb listings on the touristification of the economic base of both central and non-central neighbourhoods. Furthermore, we also attempt to investigate claims that Airbnb listings lead to above-average increases in property prices—due to investment opportunities provided—potentially causing further gentrification effects in neighbourhoods.

7.4   Research Methodology

7.4.1   Description of the Case Study

To investigate neighbourhood touristification and Airbnb effects on housing prices, the city of Amsterdam (the Netherlands) was chosen as a case study. The tourism situation in Amsterdam has received widespread research attention already, since the city exhibits strong growth patterns within a limited spatial context and since strategies to disperse tourists to the urban fringe have so far proven limited in success due to the absolute magnitude of new visitors arriving on a yearly basis. Over a 20-year period, between 2000 and 2019, the number of hotel arrivals has increased by 129.1%, from 4,015,000 in 2000 to 9,200,000 in 2019. In terms of overnight stays, this translated to 7,766,000 nights in 2000 compared to 16,943,000 nights in 2019. While the vast majority of hotel nights consists of foreign visitors, the percentage of domestic visitor nights did increase from 11.9% in 2000 to 16.9% in 2019. In its overview report on tourism, the city notes how tourism growth in Amsterdam seems to be driven by increases in accommodation supply—given that the average hotel occupancy rates is rather stable at a high 84%. The numbers of hotels increasing from 331 in 2002 to 524 in 2019; an increase in hotel bed capacity from 36,910 to 81,257 (i.e. +120.1%) (Gemeente Amsterdam 2019, 2020a).

Apart from hotel accommodation, Amsterdam has also seen a strong rise in Airbnb listings, to the extent that multiple regulatory changes were introduced in order to try to limit and contain private tourism rentals. Since 1 October 2017, Airbnb hosts are required to register at the municipality, and since 1 January 2019 the maximum number of available days is set at 30 days per year. In 2018 the city estimated the number of unique Airbnb listings at approximately 29,000—with an estimated additional 7250 apartments and houses being rented out through other platforms. Airbnb accommodations would therefore account for 7% of the housing supply—even though as per regulation, the Airbnb listings ought to still primarily function as residential property as well, given the 30 day per year maximum. While the number of guests and guest nights in Airbnb properties are unknown, the city estimated that 800,000 guests had spent 2,700,000 nights in Airbnb listings in 2018, an increase by 8% (Gemeente Amsterdam 2019).

In terms of tourism intensity, Amsterdam counted per day on average 5.3 hotel guests per 100 residents. While this intensity is already among the higher echelons of European tourism cities (comparable with Lisbon and Bruges), spatial concentration significantly influences this ratio. In Amsterdam Centre, the tourist intensity increases to 21 hotel guests per 100 residents, or a tourist footprint of 2890 guests per km^2 (Gemeente Amsterdam 2019).

7.4.2 Description of Datasets

Two secondary datasets are used in this research. From the municipality of Amsterdam, aggregated statistics on neighbourhood level, comprising 99 neighbourhoods in total, are collected for the period 2015–2019 (Gemeente Amsterdam 2020b). These data comprise over 500 variables on demographic, social, spatial, and economic characteristics of these neighbourhoods. The large dataset is trimmed down to a selection of variables that are considered particularly useful for our study, partly following the example of Lazrak et al. (2014) in order to prevent omitted variable bias.

In terms of demographic variables, we collected data on population density and percentage of the population with a non-Western migration background. While it would be interesting to also account for education levels across neighbourhoods, this data was not available for the entire time period 2015–2019 and was therefore omitted. Next, a number of neighbourhood dwelling characteristics were taken into account. First of all, the housing stock—i.e. addresses registered with the purpose 'home'—was counted per neighbourhood. Next, the percentage of this housing stock with a living space below 40 m^2, and the percentage with a living space above 100 m^2 were used to distinguish between small and large apartments and houses. Similarly, the percentage of housing stock registered as private property of the occupant—as opposed to rental properties or corporation-owned properties—was collected as an indicator of private property market availability. Finally, the average house value per m^2—determined by the office of municipal taxes and based on real local transaction data—serves as a proxy of registered sales price. These values were corrected for inflation with index $2015 = 100$. Additional neighbourhood characteristics that were obtained relate to the number of educational facilities, and number of cultural facilities.

As a useful indicator for the economic transformation of the economic fabric of the neighbourhood—i.e. the touristification of the neighbourhood—the percentage of tourist enterprises was calculated on the total number of registered companies of the neighbourhood. Tourism activities included in this variable are bars, restaurants, passenger transport, travel agencies, culture and recreation, marina and sailing, and recreational retail. Importantly, we excluded accommodation and treated number of hotel rooms in the neighbourhoods as a separate indicator (and potential driver of touristification).

The second dataset of use collected Airbnb data from the Inside Airbnb project (Inside Airbnb 2020), which scrapes the Airbnb websites at multiple time points to identify availability of listings. Since listing availability fluctuates, for each year in the 2015–2019 period, three data points were scraped: in April, August, and December. By mapping the coordinates of listings onto neighbourhood polygons, the total number of hosts and listings per neighbourhood could be calculated. A first variable collected pertains to total scraped listings, counted on a yearly basis as the mean of listings in April, August, and December. Via the same procedure, for all these listings, total accommodation capacity was calculated. To also make an estimate on overnight stays per booked listing, total yearly reviews were scraped. Finally, one limitation of working with scraped data is that not all listings that will be withheld are necessarily active. Particularly, given the '30-day maximum' regulation that was introduced in 2019, the mere total of scraped listings might overestimate the impact of Airbnb. By taking the mean of the three data points, a yearly average was calculated as all listings that had received a review within the last 90 days. Again, to come to a yearly calculation, the mean of this variable was taken over the three monthly periods.

7.4.3 Statistical Methods

Since our dataset contains both a space (99 neighbourhoods) and a time (2015–2019) dimension, we are using panel data methods for the analysis, with a balanced panel of relatively a large cross-section ($N = 99$) and a relatively short time period ($T = 5$). Prior to estimating the proposed model, the data will be tested for stationarity. A time series is considered stationary if the autocovariances of the series are unrelated to the time-dimension (Lee and Brahmasrene 2013). Not satisfying this condition could lead to spurious inference (Balaguer and Cantavella-Jordá 2002). The augmented Dickey-Fuller (ADF) unit root test checks for stochastic trend, accounting for the possibility that v_{nt} are correlated, with the null hypothesis being that the series has a unit root.

Another potential issue that is specifically related to the cross-sectional nature of panel data is existence of cross-sectional dependence or heteroskedasticity. This implies a correlation of residuals across cross-sectional units as a result of spatial or spillover effects, or underlying common factors (Baltagi and Pesaran 2007). In regional analyses, such dependencies are not uncommon. Baltagi (2013) notes how heteroskedasticity under the assumption of homoskedastic disturbances will still lead to consistent coefficient estimates, but standard errors will be biased. The Pesaran CD test of independence will be used as a test for uncorrelated residuals, since it has been shown to perform better in cases with small T and large N. Finally, another classic assumption about v_{nt} is that correlation among error terms is only due to repeated measurement of the same cross-sectional units. However, this might be too restrictive an assumption for economic variables where fluctuations often span multiple periods (Baltagi 2013). The Breusch-Godfrey/Wooldridge test for serial correlation will be used to test for serial correlation.

Our standard statistical approach will compare a simple pooled regression model, with constant coefficients across time and neighbourhoods, with fixed effects models with cross-sectional variance and time invariance (Eq. 7.1), cross-sectional invariance and time variance (Eq. 7.2), and both cross-sectional and time variance (Eq. 7.3):

$$Y_{nt} = \lambda_{1n} + \sum_{i=1} \beta_i X_{\text{int}} + v_{nt} \tag{7.1}$$

$$Y_{nt} = \lambda_{2t} + \sum_{i=1} \beta_i X_{\text{int}} + v_{nt} \tag{7.2}$$

$$Y_{nt} = \lambda_{1n} + \lambda_{2t} + \sum_{i=1} \beta_i X_{\text{int}} + v_{nt} \tag{7.3}$$

with n = neighbourhoods, t = years, i = number of explanatory variables, and X = a vector of explanatory variable scores. Equation (7.1) shows that the intercept λ is neighbourhood-dependent and can vary for each cross-section—estimated via dummy-variables—, while the β coefficients of the explanatory variables are considered constant across time t and cross-section n. Equation (7.2) introduces dummy-variables for each time-section t, leading to year-specific λ-estimates, while removing neighbourhood-specific intercepts. Equation (7.3), finally, mixes both approaches by allowing for varying time and neighbourhood intercepts. In all cases, the error term follows the classical assumptions that $E(v_{nt}) \sim N(0, \sigma^2)$.

Depending on the results of the test statistics for serial correlation and heteroskedasticity in disturbances, a robust Panel-Corrected Standard Errors (PCSE) model might be used, following the suggestion of Moundigbaye et al. (2017). In such case, the error term in Eqs. (7.1–7.3) will be written as:

$$v_{nt} = \mu_n + \nu_{nt} \qquad\qquad (7.4)$$

with the error term consisting of μ_n, denoting a time-invariant and cross-sectional-specific fixed parameter to be estimated, and a remaining disturbance ν_{nt} with an IID $(0, \sigma^2{}_\nu)$ distribution.

7.5 Results

7.5.1 Explorative Data on Airbnb Listings in Amsterdam

Before assessing the relationship between Airbnb listings on neighbourhood touristification and housing prices, a short overview of main trends in Airbnb for Amsterdam as a whole is given in Table 7.1. Since these listings are variably offered, collecting a stable population is challenging. In order to improve reliability and exempt listings that were seemingly inactive, yet still present on the website, listings without any reviews were removed. This led to roughly 2500 (slightly varying in each period) listings being dropped in each monthly dataset. For listings that can be observed and are found to be active (i.e. at least one review), it needs to further be examined whether there was any recent customer interaction, since even active

Table 7.1 Trends in Airbnb supply (2015–2019)

Year	Month	Airbnb listings	Capacity	Active listings	Total reviews
2015	Yearly avg./total	7435	22,537	5511	84,231
	April	5991	18,506	3583	
	August	7376	22,297	6150	
	December	8922	26,732	6788	
2016	Yearly avg./total	11,326	33,440	7801	64,235
	April	8927	26,509	5496	
	August	11,986	35,618	9182	
	December	13,087	38,252	8742	
2017	Yearly avg./total	14,717	42,346	8518	49,516
	April	12,815	37,061	6636	
	August	15,445	44,373	10,363	
	December	15,941	45,767	8592	
2018	Yearly avg./total	16,916	48,495	8237	42,094
	April	16,068	46,152	6962	
	August	17,438	49,938	9676	
	December	17,394	49,844	8180	
2019	Yearly avg./total	17,301	49,252	7189	35,096
	April	16,994	48,488	6040	
	August	17,833	50,806	8581	
	December	17,250	48,956	7048	

listings are not continuously available. To this extent, only listings were counted that had received reviews over the past 90 days.

Table 7.1 shows a continuous rise in Airbnb listings in Amsterdam from—on average—7435 in 2015 to 17,301 in 2019. As the monthly data shows, the population is not stable, with April counting less-observed listings than August and December. Our estimate of Airbnb listings is notably smaller than the assessment made by the municipality of Amsterdam, who estimated approximately 29,000 listings in 2018 (Gemeente Amsterdam 2019). This might be explained by the fact that the municipality makes the yearly sum of all listings that had at least 1 day of availability during that calendar year. In comparison, we take a more conservative approach of averaging active listings across three time points in each year.

Together with listing numbers, logically also the capacity provided in Airbnb properties across Amsterdam has risen significantly from approximately 22,537 beds in 2015 to 49,252 beds in 2019. With a hotel bed capacity in 2019 of 81,257, we can therefore estimate that the lodging capacity provided by Airbnb is now more than half of the commercial hotel accommodation space. It has to be taken into account that while hotel beds are usually offered all year round, Airbnb listings might only be available for limited periods throughout the year. This is particularly relevant considering new regulations since 1 January 2019 which limit the possibility of renting out Airbnb listings to a maximum of 30 rental days per year.

The data on active listings (i.e. listings that received a guest review less than 90 days prior) shows that not all scraped listings were likely receiving guests recently—albeit the reliability of this variable is dependent on the proclivity of guests to write reviews. In this sense we can see that while absolute Airbnb listings show a rising trend, percentage-wise, a lower amount of those listings seem to be continuously active. In 2015 74.1% of listings had received a review in the last 90 days. In 2017, this percentage was 57.9%, while in 2019 the percentage had dropped to 41.6%. This might be explained by a combination of rising Airbnb supply and competition and the effects of policy regulation. At the same time, however, we cannot exclude the possibility that these data merely reflect a change in consumer behaviour with contemporary tourists potentially being less inclined to write reviews.

Next, a spatial analysis seeks to investigate the distribution of Airbnb listings across Amsterdam neighbourhoods. The map in Fig. 7.1 (left panel) is based on the yearly average of the 2019 data. As could be expected, listings are more prevalent in the historic central district, in and around main tourist areas such as *Museumplein*, *Vondelpark*, *Dam*, and *Jordaan*. Seven neighbourhoods in Amsterdam counted over 400 accommodations: *Jordaan* (834), *Oude Pijp* (803), *Nieuwe Pijp* (592), *Landlust* (579), *Staatsliedenbuurt* (522), *Indische Buurt West* (417), and *Frederik Hendrikbuurt* (401). At the other end of the spectrum, six neighbourhoods had less than ten Airbnb listings: *Amstel III/Bullewijk* (1), *Westelijk Havengebied* (1), *Noordelijke IJ-oevers Oost* (5), *Driemond* (6), *Lutkemeer/Ookmeer* (8), and *Sloterdijk* (8).

In order to map the listing evolution, Fig. 7.1 (right panel) shows, in absolute terms, the growth in the number of listings between 2015 and 2019, at local level.

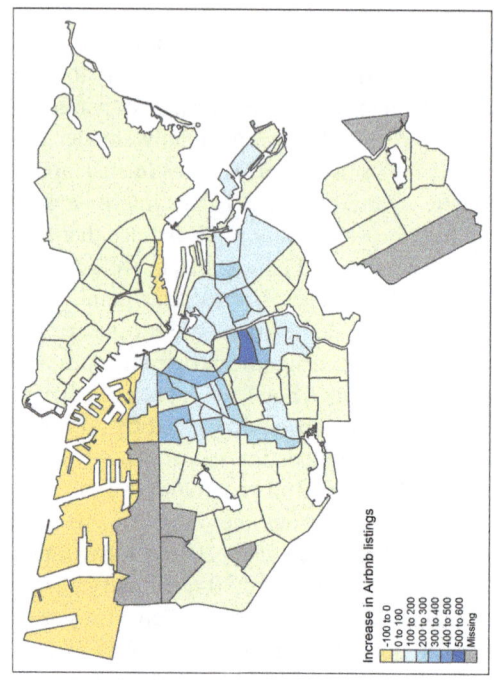

Increase in Airbnb listings

-100 to 0
0 to 100
100 to 200
200 to 300
300 to 400
400 to 500
500 to 600
Missing

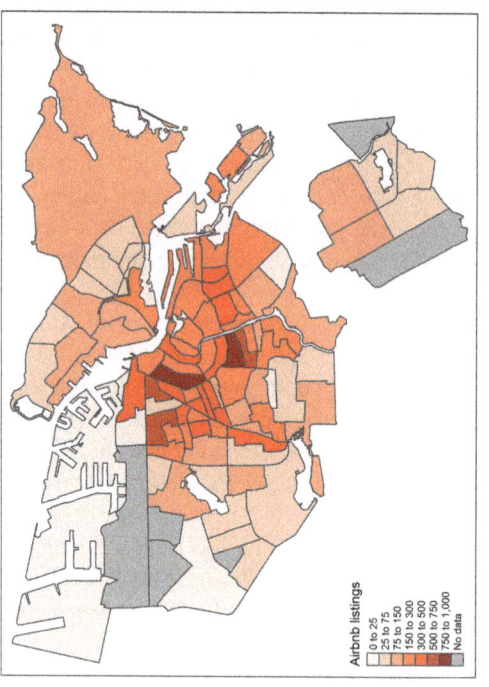

Airbnb listings

0 to 25
25 to 75
75 to 150
150 to 300
300 to 500
500 to 750
750 to 1,000
No data

Fig. 7.1 Airbnb listings—2019 (left) and growth in Airbnb listings—2015–2019 (right)

Only four neighbourhoods—largely overlapping with the areas where Airbnb listings are scarce—showed a decrease: *Houthavens* (-11), *Noordelijke IJ-oevers Oost* (-5), *Westelijk Havengebied* (-4), and *Sloterdijk* (-1). In the vast majority of cases, Airbnb listings grew significantly, with 36 neighbourhoods showing an increase of over 100 listings. The main growth poles correlate with areas that already had a strong presence of accommodation services to begin with: *Oude Pijp* grew by 501 listings, *Landlust* by 373, *Nieuwe Pijp* by 352, *Jordaan* by 334, and *Staatsliedenbuurt* by 326.

7.5.2 Airbnb as a Driver of Rising Housing Costs and Neighbourhood Touristification

The first part of the analysis links Airbnb proximity to residential opportunities by investigating whether an increase in Airbnb listings is statistically related to rising housing prices in Amsterdam neighbourhoods. The general principle of Lazrak et al. (2014) is followed in the housing price function, by including transactional, structural, and spatial characteristics to the extent that such data was available on neighbourhood level. Adopting a log-linear transformation, the natural log of house value per m^2 was regressed against population density, percentage of the population with a non-Western migration background, housing stock, percentage of private ownership, percentage with living space below $40\,m^2$, percentage with living space above $100\,m^2$, number of educational facilities, number of cultural facilities, and Airbnb listings.

As stated by Baltagi (2013), in micro-panels with relatively short time series, variables are more likely to exhibit stationary stochastic processes and the ADF unit root test indeed confirmed the stationarity of our data on 2 and 4 lags, with p-values <0.01. Results for the Pesaran CD test of independence ($Z = 52.162$; p-value <0.001) lead us to reject the null hypothesis that residuals are uncorrelated. Unobserved factors might therefore, affect the model variables differently for various neighbourhoods across Amsterdam. The Breusch-Godfrey/Wooldridge test for serial correlation further rejects the null hypothesis of no serial correlation ($\chi^2 = 55.381$; d$f = 1$; p-value <0.001), potentially resulting in consistent but inefficient coefficient estimates and biased standard errors if not properly accounted for. Given these findings, the fixed effects model with robust PCSE estimates was considered most appropriate.

As mentioned in the methodology, apart from the standard pooled model, the fixed effects model was estimated for area-variant intercepts (Eq. 7.1), time-variant intercepts (Eq. 7.2), and two-way intercepts (Eq. 7.3). Due to the construction of the panel with a relative large N and small T, including area-variant intercepts (in Eqs. 7.1 and 7.3) for all neighbourhoods created computational issues. Therefore, a simplified least-squares dummy variable (LSDV) model was used that created area-variant intercepts on the level of eight Amsterdam regions (combining multiple

Table 7.2 Results of two-way fixed effects PCSE model, y = housing value per m^2

	Estimate	Std. error	t-value	$Pr(>|t|)$
Intercept	8.30790***	0.04976	166.962	<0.001
City area B: Westpoort	−0.82315***	0.05397	−15.251	<0.001
City area E: West	−0.05304°	0.02838	−1.869	0.062
City area F: Nieuw-west	−0.30413***	0.04208	−7.228	0.000
City area K: Zuid	0.00619	0.03241	0.191	0.849
City area M: Oost	−0.11843**	0.03828	−3.094	0.002
City area N: Noord	−0.32561***	0.03606	−9.029	<0.001
City area T: Zuidoost	−0.48730***	0.07656	−6.365	<0.001
Year 2016	0.05065***	0.00539	9.397	<0.001
Year 2017	0.18845***	0.01334	14.128	<0.001
Year 2018	0.31775***	0.01599	19.868	<0.001
Year 2019	0.39732***	0.01630	24.371	<0.001
Population density	0.00000**	0.00000	3.061	0.002
% of non-European migrants	−0.00546***	0.00077	−7.045	<0.001
Housing stock	−0.00001	0.00001	−1.497	0.135
% of private ownership	−0.00221°	0.00117	−1.889	0.060
% living area < 40 m^2	0.00005	0.00107	0.050	0.960
% living area > 100 m^2	0.00237*	0.00102	2.323	0.021
Educational facilities	−0.00093	0.00089	−1.046	0.296
Cultural facilities	0.00039°	0.00021	1.842	0.066
Airbnb listings	0.00025*	0.00011	2.267	0.024
Adjusted R^2 = 0.911 F-statistic = 239.351 on DF 20 and 445 (p-value < 0.001) Total sum of squares = 51.619 Residual sum of squares = 4.3903				

Note: ***≤0.001, **≤0.01, *≤0.05, °≤0.1

adjacent neighbourhoods). F tests for individual and/or time effects were used in order to select the superior model, with the two-way variant fixed effects model (Eq. 7.3) proving significantly better than the other alternatives. Table 7.2 gives an overview of the results of this panel-based error correction model using two-way, individual and, time-variant fixed effects.

First of all, the significant nature of nearly all intercept dummies for areas and years indicates that some unobserved effects are correlated with explanatory variables. The fact that most areas—apart from the non-significant coefficient for *Zuid*—exhibit negative coefficients relates to the differential nature of the dummy variables, which compare the intercepts of other areas with the *Centre* (i.e. City area A), whose value is included in the standard intercept. Since the *Centre* is—together with *Zuid*—the highest priced area in Amsterdam, other regions compare negatively, with *Westpoort* being the area where average housing price per m^2 is lowest, followed by *Zuidoost*. The significant positive yearly dummy variables imply that the average housing price per m^2 across Amsterdam has been rising faster than inflation—since the average housing price had been corrected for inflation prior to its model inclusion.

In terms of explanatory variables, across the panel there seems to be a positive influence of local characteristics such as population density (4.929e-06; p-value $= 0.002$), while neighbourhoods where larger proportions of residents are of non-European migration background exhibit significantly lower housing values (-5.46e-03; p-value < 0.001). Neighbourhoods with higher proportions of large houses and apartments above 100 m^2 are logically also showing higher housing values (2.372e-03; p-value $= 0.021$). Of interest to the scope of our research is the positive and significant coefficient of Airbnb listings (2.484e-04; p-value $= 0.024$), indicating that the amount of Airbnb listings in a neighbourhood inflated the property value and selling price in the 2015–2019 period. While such an effect can be considered positive for home owners, the threshold for potential home buyers increases, potentially contributing to a gentrification of neighbourhoods.

Figure 7.2 maps both the housing value per m^2 for the 2019 data (left panel) and the increase in housing value in absolute terms between 2015 and 2019 (right panel) to allow for a spatial consideration and comparison to Fig. 7.1. Quite logically, the central area of Amsterdam within the *Singel* and running South towards *Museumplein* and *Vondelpark* are neighbourhoods with the highest property values. Since these districts are rich in tourism experiences, it may be seen from Fig. 7.1 that they also accounted for the highest numbers of Airbnb listings and fastest growth. For the four neighbourhoods with fastest rising housing values (between +€2500 and +€3000 per m^2), a contrasting presence of Airbnb listings can be noted. On the one hand, *Omval/Overamstel* (housing value $= +€2909$ and Airbnb listings $= +86$) and *Zeeburgereiland/Nieuwe diep* (housing value $= +2669$ and Airbnb listings $= +46$) showed strong increases on both variables, while on the other hand *Sloterdijk* (housing value $= +€2515$ and Airbnb listings $= -1$) and *Noordelijke IJ-oevers Oost* (housing value $= +€2702$ and Airbnb listings $= -5$) experienced a strong rise in property value without any notable presence of Airbnb listings.

While the results seem to suggest that Airbnb has affected property values across Amsterdam, our primary interest is linked to its effect on neighbourhood characteristics in terms of potentially transforming the economic base of local areas towards tourism-supporting functions, thereby crowding-out residential-specific activities. Ultimately, such transformations are driven by an increase in end-users (i.e. tourists), which can take the form of both day tourism and overnight tourism. In other words, a particular neighbourhood might also experience a transformation without a direct rise in local accommodation opportunities, if day tourists start to frequent the area. Unfortunately, there are no stable methods to consistently measure day tourists on local neighbourhood level; therefore, the neighbourhood touristification function will only include accommodation-specific data and regresses the percentage of tourist enterprises—including bars, restaurants, passenger transport, travel agencies, culture and recreation, marina and sailing, and recreational retail—against the number of hotel rooms and the number of Airbnb listings.

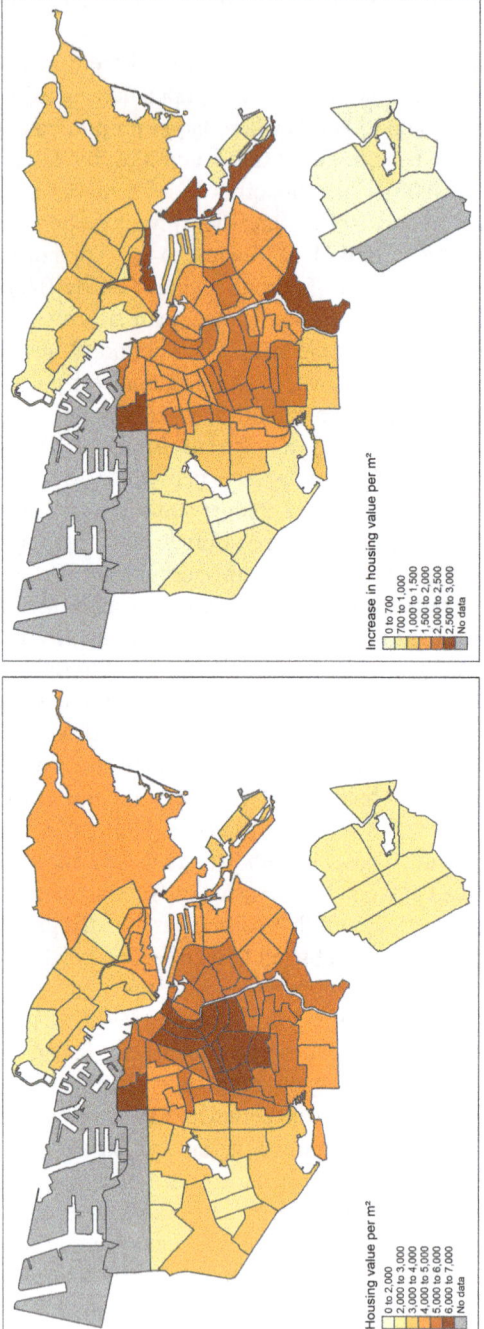

Fig. 7.2 Housing value—2019 (left) and growth in housing value—2015–2019 (right)

Table 7.3 Results of one-way fixed effects PCSE model, y = touristification index

	Estimate	Std. error	t-value	Pr($>|t|$)
Intercept	6.55639***	1.60635	4.082	<0.001
City area B: Westpoort	−2.66278°	1.52489	−1.746	0.082
City area E: West	−1.19405	1.47380	−0.810	0.418
City area F: Nieuw-west	3.50393	2.24199	1.563	0.119
City area K: Zuid	−3.23704*	1.43819	−2.251	0.025
City area M: Oost	−2.23978	1.42353	−1.573	0.117
City area N: Noord	1.49507	2.04766	0.730	0.466
City area T: Zuidoost	−0.59463	1.92632	−0.309	0.758
Number of hotel beds	0.00111**	0.00040	2.778	0.006
Airbnb listings	0.00613**	0.00218	2.809	0.005
Adjusted R^2 = 0.431 F-statistic = 26.512 on DF 9 and 294 (p-value < 0.001) Total sum of squares = 5057.4 Residual sum of squares = 2791.7				

Note: ***≤0.001, **≤0.01, *≤0.05, ≤0.1

Once more, the ADF unit root test was adopted and confirmed the stationarity on both 2 and 4 lags with p-values <0.01. The Pesaran CD test of independence (Z = 8.6234; p-value <0.001) and the Breusch-Godfrey/Woolridge test for serial correlation (χ^2 = 18.459; df = 1; p-value <0.001) show comparable results to our earlier housing price analysis and are indicative of serial correlation and heteroscedasticity in error terms, leading us to adopt fixed effects model with robust PCSE estimates. After testing the standard pooled model and fixed effects models with area and/or time-variant intercepts (Eq. 7.1–7.3)—once again using area-variant intercepts on the aggregated level of 8 Amsterdam regions—and comparing model results with the F Test, the one-way area variant fixed effects was selected. Different from the housing price analysis, in the touristification-equation, none of the time-varying dummies were significant, indicating that the year in itself did not correlate with unobserved effects. Table 7.3 gives an overview of the results of the panel-based error correction model using one-way, time-invariant fixed effects.

From the analysis of city region dummy intercepts, we can conclude that the level of touristification is quite similar across the *Centre* (City area A)—i.e. the standard intercept—and the areas *West, Nieuw-West, Oost, Noord*, and *Zuidoost*—all having non-significant differential intercepts. While the adjusted R^2 (0.431) is lower, compared to the housing price model, due to a more limited number of explanatory variables, both model variables are positively related to an increase in neighbourhood touristification. We can therefore conclude that an increase in overnight tourism accommodations, both commercially in terms of hotel beds (1.108e-03; p-value = 0.006) and via peer-to-peer networks in terms of Airbnb listings (6.128e-03; p-value = 0.005), leads to significant changes in the economic fabric of the neighbourhood and a replacement of urban functionalities towards tourism.

Figure 7.3 maps the 2019 data on the percentage of tourism-related activities (excluding accommodation) on neighbourhood level (left panel) and also compares relative growth or decline (in percentage points) in tourism-related activities between 2015 and 2019 (right panel). Visually, a comparison can be made to Fig. 7.1 in order to identify spatial links with Airbnb listings. The strongest presence of tourism industries is naturally found around the central axis. In the *Centre* (City area A), in four neighbourhoods the number of tourism-related businesses excluding accommodation is at least 15% of total economic activities: *Burgwallen-Nieuwe Zijde* (24%), *Burgwallen-Oude Zijde* (24%), *Grachtengorden West* (17%), and *De Weteringschans* (16%). In *Nieuw-West* (City area F) another five neighbourhoods count at least 15% of tourism-related businesses in their total: *Geuzenveld* (18%), *Slotermeer-Zuidwest* (17%), *Osdorp-Midden* (17%), *Slotermeer-Noordoost* (15%), and *Eendracht* (15%). Finally, one neighbourhood in *West* (*De Kolenkit*, 16%) and one area in Noord (*Waterlandpleinbuurt*, 18%) exhibited high tourism-related activity patterns.

From Fig. 7.3 (right panel), we can see that the growth in tourism-related activities in marginal in the central district, which is probably already quite satiated. Instead, growth seems to be spread quite evenly around secondary neighbourhoods outside of the *Singel*, with a rise in about 2–6% points. Similar to the growth in Airbnb listings, the increase in tourism industries—albeit at relatively slow pace—is a reality in practically all neighbourhoods of the city. Albeit that the growth in Airbnb accommodation seemed a bit more concentrated in the South and Southeast area of the central city, while the touristification ratio grows fastest in the West and North areas.

7.6   Discussion and Conclusion

The strong growth in urban tourism over the past decades has contributed to a situation where an ever-increasing number of both day and overnight visitors need to be accommodated. Limited by natural spatial constraints, cities where tourism was historically concentrated in tourism precincts have increasingly witnessed a spread of visitors towards less central, more residential areas, slowly transforming the entire city into a tourism precinct. This trend has partly been caused by policy measures designed to spread tourism and decrease the pressure on the certain central neighbourhoods, as well as changing tourism experiences and motivations increasingly looking for 'off-the-beaten-track' locations and opportunities to 'live like locals'. Such trend has further been supported by a growth in peer-to-peer networks such as Airbnb that markets its listings as an opportunity to experience a location in a more authentic environment. As such, Airbnb has recently generated much media attention, with various destinations beginning to enforce stricter regulation, primarily out of a concern that an increased Airbnb presence will further diminish an already insufficient housing stock, thereby raising prices for local residents and

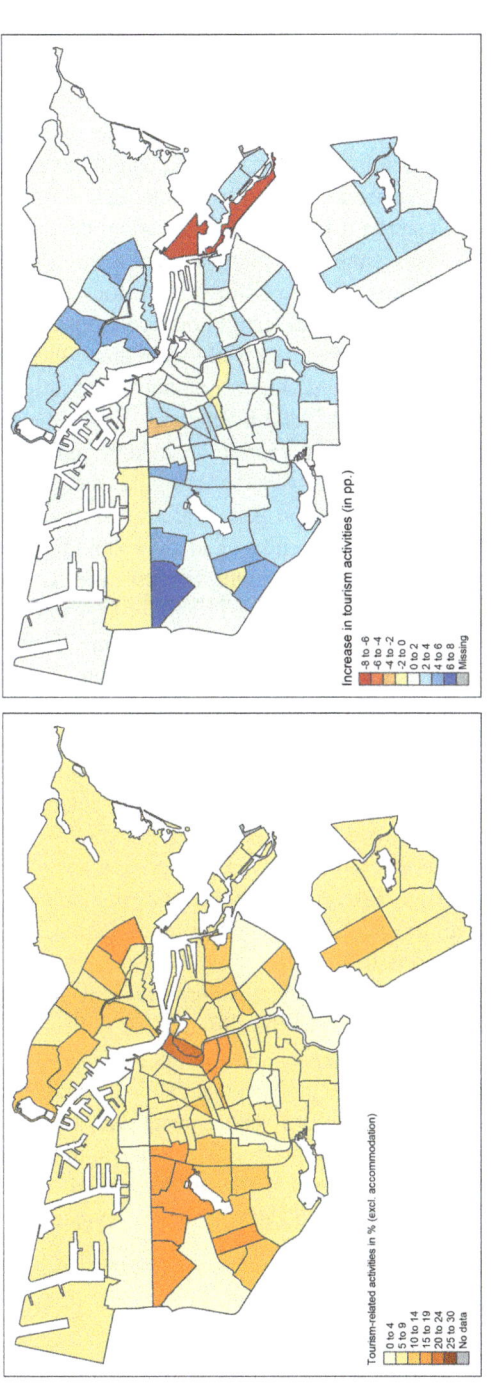

Fig. 7.3 Tourism-related activities (excl. accommodation) in %—2019 (left) and growth in tourism-related activities (excl. accommodation) in percentage points—2015–2019 (right)

potentially leading to the creation of gentrified tourist areas which will then further serve a crowding out of local functionalities.

Based on a supply-sided database of Airbnb listings for the period 2015–2019, our study aimed to find effects of Airbnb on housing values—as a proxy for sales prices—and on a touristification index—modelled as the percentage of tourism-related activities in total economic activities within a neighbourhood, and excluding accommodation. Combining Airbnb data with various neighbourhood characteristics collected from the Amsterdam municipality, the final dataset spanned 5 years and 99 neighbourhoods. This panel was analysed via a fixed effects model with Panel-Corrected Standard Errors. Airbnb listings were found to have a significant positive effect on both housing values and neighbourhood touristification, indicating that, ceteris paribus, Airbnb listing presence is higher in areas where the selling price of property is higher and in locations where non-accommodation tourism-related industries take up a larger portion of total businesses. While the effects are found significant, the direction of causation cannot be readily inferred. So it remains unclear whether Airbnb listings are a driver for neighbourhood touristification and potential gentrification through rising property values, or whether Airbnb listings follow neighbourhood touristification trends—and are potentially used as an income generator when an expensive property is bought. Furthermore, the models also make clear that Airbnb listings are not solely responsible for changes in the urban fabric of neighbourhoods. While the model on touristification was limited by a lack of neighbourhood-specific day tourism data, commercial hotel developments had a similar positive effect as Airbnb listings. Therefore, while questions might be asked about Airbnb and a lack of manageability, it seems too straightforward to blame overtourism issues on the platform when, at the same time, Amsterdam allowed the number of hotel beds to increase by 120.1% between 2002 and 2019. Touristification of neighbourhoods and protecting residential environments is a complex problem that requires a multi-layered approach by not only regulating peer-to-peer networks, but also limiting commercial accommodation supply, spreading tourists across the hinterland, and managing international arrivals at point of entry (e.g., capping flight capacity).

Panel time series research on the subject is currently still limited by the relative novelty of the phenomenon, as well as the difficulty in collecting information on Airbnb listings. As a result, time series are still short and future research might gradually become more intricate, when additional data collection allows for it. Such future research might consider the use of spatial panel data analysis which can test for the existence of spatial interaction and spillover effects. In this context, also the complex competitive interdependency between the hotel sector and Airbnb (a similar) platforms would have to be thoroughly investigated at neighbourhood level.

References

Airbnb (2019) New 2018 data: Airbnb grows responsibly and spreads tourism across the country. https://press.airbnb.com/new-2018-data-airbnb-grows-responsibly-and-spreads-tourism-across-the-country/. Accessed 4 Sept 2019

Asakura Y, Iryo T (2007) Analysis of tourist behaviour based on the tracking data collected using a mobile communication instrument. Transport Res A Pol Pract 41(7):684–690

Ashworth GJ, Tunbridge JE (2000) The tourist-historic city: retrospect and prospect of managing the heritage city. Elsevier Science, Oxford

Balaguer J, Cantavella-Jordá M (2002) Tourism as a long-run economic growth factor: the Spanish case. Appl Econ 34(7):877–884

Baltagi BH (2013) Econometric analysis of panel data, 5th edn. Wiley, Chichester

Baltagi BH, Pesaran H (2007) Heterogeneity and cross section dependence in panel data models: theory and applications. J Appl Econ 22:229–232

Barrera-Fernández D, Garcia Bujalance S, Scalici M (2019) Touristification in historic cities: reflections on Malaga. Rev Tur Contemp 7(1):93–115

Business Insider (2016) Here's exactly what Airbnb does to rent in popular cities. https://www.businessinsider.com/statistics-data-airbnb-rent-prices-2016-10?r=US&IR=T. Accessed 4 Sept 2019

Choi KH, Jung JH, Ryu SY, Do Kim S, Yoon SM (2015) The relationship between Airbnb and the hotel revenue: in the case of Korea. Indian J Sci Technol 8(26):1–8

du Cros H (2007) Too much of a good thing? Visitor congestion management issues for popular world heritage tourist attractions. J Herit Tour 2(3):225–238

Daskalopoulou I, Petrou A (2009) Urban tourism competitiveness: networks and the regional asset base. Urban Stud 46(4):779–801

Dodds R (2007) Sustainable tourism and policy implementation: lessons from the case of Calviá, Spain. Curr Issues Tour 10(4):296–322

Fainstein SS (2008) Mega-projects in New York, London and Amsterdam. Int J Urban Reg 32 (4):768–785

Fava N, Rubio SP (2017) From Barcelona: the pearl of the Mediterranean to bye bye Barcelona. In: Bellini N, Pasquinelli C (eds) Tourism in the city. Springer, New York, pp 285–295

Freytag T, Bauder M (2018) Bottom-up touristification and urban transformations in Paris. Tour Geogr 20(3):443–460

Füller H, Michel B (2014) 'Stop being a tourist!' New dynamics of urban tourism in Berlin-Kreuzberg. Int J Urban Reg Res 38(4):1304–1318

Fusco Girard L, Nijkamp P (eds) (2009) Cultural tourism and sustainable local development. Ashgate, Aldershot

Gemeente Amsterdam (2019) Toerisme MRA 2018-2019. Onderzoek, Informatie en Statistiek, Amsterdam

Gemeente Amsterdam (2020a) Dossier Toerisme. https://data.amsterdam.nl/dossiers/dossier/toerisme/fdcc54a1-5aa7-4ddf-af16-1c28a99b8c5f/. Accessed 13 Aug 2020

Gemeente Amsterdam (2020b) Basisbestand Gebieden Amsterdam (BBGA). https://data.amsterdam.nl/datasets/G5JpqNbhweXZSw/basisbestand-gebieden-amsterdam-bbga/. Accessed 2 Aug 2020

Giaoutzi M, Nijkamp P (eds) (2017) Tourism and regional development. Routledge, London

Gilbert D, Clark M (1997) An exploratory examination of urban tourism impact, with reference to residents attitudes, in the cities of Canterbury and Guildford. Cities 14(6):343–352

Glaeser E, Kourtit K, Nijkamp P (eds) (2020) Urban empires, cities as global rulers in the new urban world. Routledge, London

Gravari-Barbas M, Guinand S (eds) (2017) Tourism and gentrification in contemporary metropolises: international perspectives. Routledge, New York

Gravari-Barbas M, Jacquot S (2016) No conflict? Discourses and management of tourism-related tensions in Paris. In: Colomb C, Novy J (eds) Protest and resistance in the tourist city. Routledge, London, pp 31–51

Gursoy D, Rutherford DG (2004) Host attitudes toward tourism: an improved structural model. Ann Tour Res 31(3):495–516

Gutiérrez J, García-Palomares JC, Romanillos G, Salas-Olmedo MH (2017) The eruption of Airbnb in tourist cities: comparing spatial patterns of hotels and peer-to-peer accommodation in Barcelona. Tour Manage 62:278–291

Hall CM, Page SJ (2009) Progress in tourism management: from the geography of tourism to geographies of tourism – a review. Tour Manage 30:3–16

Hasegawa H (2010) Analyzing tourist satisfaction: a multivariate ordered probit approach. Tour Manage 31(1):86–97

Heo CY, Blal I, Choi M (2019) What is happening in Paris? Airbnb, hotels, and the Parisian market: a case study. Tour Manage 70:78–88

Inside Airbnb (2020) Get the data. http://insideairbnb.com/get-the-data.html. Accessed 2 Aug 2020

Ioannides D, Röslmaier M, van der Zee E (2019) Airbnb as an instigator of 'tourism bubble' expansion in Utrecht's Lombok neighbourhood. Tour Geogr 21(5):822–840

Jansen-Verbeke M (1986) Inner-city tourism: resources, tourists and promoters. Ann Tour Res 13(1):79–100

Jansen-Verbeke M, Lievois E (1999) Analysing heritage resources for urban tourism in European cities. In: Pearce DG, Butler R (eds) Contemporary issues in tourism development. Routledge, London, pp 81–107

Kadi K, Plank L, Seidl R (2019) Airbnb as a tool for inclusive tourism? Tour Geogr. https://doi.org/10.1080/14616688.2019.1654541

Kelly I (2008) Urban tourism precincts: an overview of key themes and issues. In: Hayllar B, Griffin T, Edwards D (eds) City spaces – tourist places: urban tourism precincts. Elsevier, Oxford, pp 107–126

Keul A, Kühberger A (1997) Tracking the Salzburg tourist. Ann Tour Res 24(4):1008–1012

Koens K, Postma A, Papp B (2018) Is overtourism overused? Understanding the impact of tourism in a city context. Sustainability 10(12):4384

Law CM (1992) Urban tourism and its contribution to economic regeneration. Urban Stud 29(3/4):599–618

Law CM (2002) Urban tourism: the visitor economy and the growth of large cities, 2nd edn. Continuum, London

Lazrak F, Nijkamp P, Rietveld P, Rouwendal J (2014) The market value of cultural heritage in urban areas: an application of spatial hedonic pricing. J Geogr Syst 16:89–114

Lee JW, Brahmasrene T (2013) Investigating the influence of tourism on economic growth and carbon emissions: evidence from panel analysis of the European Union. Tour Manage 38:69–76

Maitland R (2006) Cultural tourism and the development of new tourism areas in London. In: Richards G (ed) Cultural tourism: global and local perspectives. Haworth Press, London, pp 113–130

Maitland R (2008) Conviviality and everyday life: the appeal of new areas of London for visitors. Int J Tour Res 10:15–25

Maitland R, Newman P (2004) Developing metropolitan tourism on the fringe of Central London. Int J Tour Res 6:339–348

Malecki E (2002) Hard and soft networks for urban competitiveness. Urban Stud 39(5/6):929–945

McKercher B (1993) The unrecognized threat to tourism: can tourism survive 'sustainability'? Tour Manage 14(1):131–136

Meethan K (2001) Tourism in global society: place, culture, consumption. Palgrave Macmillan, Basingstoke

Moundigbaye M, Rea WS, Reed WR (2017) Which panel data estimator should I use?: a corrigendum and extension. Economics 2017-58:1–38

Neuts B (2016) An econometric approach to crowding in touristic city centres: evaluating the utility effect on local residents. Tour Econ 22(5):1055–1074

Neuts B (2020) Tourism and urban economic growth: a panel analysis of German cities. Tour Econ 26(3):519–527

Nilsson JH (2020) Conceptualizing and contextualizing overtourism: the dynamics of accelerating urban tourism. Int J Tour Cities. https://doi.org/10.1108/IJTC-08-2019-0117

Novy J (2016) The selling (out) of Berlin and the de- and re-politicization of urban tourism in Europe's 'Capital of Cool'. In: Colomb C, Novy J (eds) Protest and resistance in the tourist city. Routledge, London, pp 52–72

Novy J (2018) 'Destination' Berlin revisited. From (new) tourism towards a pentagon of mobility and place consumption. Tour Geogr 20(3):418–442

Ojeda AB, Kieffer M (2020) Touristification. Empty concept or element of analysis in tourism geography? Geoforum. https://doi.org/10.1016/j.geoforum.2020.06.021

Page SJ, Hall CM (2003) Managing urban tourism. Pearson Education Limited, Harlow

Pappalepore I, Maitland R, Smith A (2014) Prosuming creative urban areas. Evidence from East London. Ann Tour Res 44:227–240

Paulauskaite D, Powell R, Andres Coca-Stefaniak J, Morrison AM (2017) Living like a local: authentic tourism experiences and the sharing economy. Int J Tour Res 19(6):619–628

Pavlic I, Portolan A, Butorac M (2013) Urban tourism towards sustainable development. Int J Multidiscip Bus Sci 1(1):72–79

Pérez EA, Nadal JR (2005) Host community perceptions: a cluster analysis. Ann Tour Res 32 (4):925–941

Sequera J, Nofre J (2018) Shaken, not stirred. New debates on touristification and the limits of gentrification. Anal Urban Change Theory Action 22(5–6):843–855

Shaw G, Williams A (1994) Critical issues in tourism: a geographical perspective. Blackwell, Oxford

Shaw S, Bagwell S, Karmowska J (2004) Ethnoscapes as spectacle: reimaging multicultural districts as new destinations for leisure and tourism consumption. Urban Stud 4:1983–2000

Shoval N (2008) Tracking technologies and urban analysis. Cities 25:21–28

Xu F, Hu M, La L, Wang J, Huang C (2020) The influence of neighbourhood environment on Airbnb: a geographically weighed regression analysis. Tour Geogr 22(1):192–209

Zervas G, Proserpio D, Byers JW (2017) The rise of the sharing economy: estimating the impact of Airbnb on the hotel industry. J Mark Res 54:687–705

Part IV
Tourism Development, Sustainability and Resilience

Chapter 8
Tourism and Economic Resilience: Implications for Regional Policies

Gabriela Carmen Pascariu, Bogdan-Constantin Ibănescu, Peter Nijkamp, and Karima Kourtit

Abstract Tourism is an important key sector in regional and national economies which appears to have often a favorable recovery potential after a shock, leading to the notion of resilience capacity of regions. In the context of a tourism-led growth mechanism, the concept of tourism-led resilience capacity is introduced (constituted of sustained tourism resilience and speed of recovery). The analytical framework is tested for the 2008–2012 financial crisis in European Union by examining relevant data from European NUTS 2 regions. The research is unfolded on two complementary axes: (a) assessing the resilience of the tourism sector, and (b) estimating the weight of tourism in the overall resilience performance of EU regions. Finally, several implications for regional and European policies are addressed as well, particularly related to the role of innovation and diversification in increasing the recovery speed following a disruption.

Keywords Tourism-led growth · Resilience · Tourism-led resilience capacity · Vulnerability · Resistance · Speedy resilience

8.1 Introduction

For decades, tourism has been considered one of the fastest growing industries in the world; due to its dynamizing effects on economic growth and job creation, it became a focal point of interest in the global economy, especially for lower income regions. In 2019, for example, its growth rate (3.5%) surpassed the growths recorded in healthcare (3.0%), retail & wholesale (2.4%), agriculture (2.3%), construction

G. C. Pascariu (✉) · B.-C. Ibănescu
Centre for European Studies, Alexandru Ioan Cuza University of Iasi, Iasi, Romania
e-mail: gcpas@uaic.ro

P. Nijkamp · K. Kourtit
Centre for European Studies, Alexandru Ioan Cuza University of Iasi, Iasi, Romania

Open University, Heerlen, The Netherlands

© The Author(s), under exclusive licence to Springer Nature Singapore Pte Ltd. 2021
S. Suzuki et al. (eds.), *Tourism and Regional Science*, New Frontiers in Regional Science: Asian Perspectives 53, https://doi.org/10.1007/978-981-16-3623-3_8

(2.1%), or the manufacturing sector (1.7%) (UNWTO 2020a), while in the previous years it managed to already overpass the growth recorded by information and communication technologies or financial services. In fact, for nine consecutive years (2011–2019), the tourism growth (3.5%) exceeded the growth of the global economy (2.5%) (UNWTO 2020a; WTTC 2020a). Its impact was substantial, with a direct relative contribution to global GDP of 3.1%, while its total contribution (direct and indirect) reached up to 10.2%. Moreover, tourism-related activities were sustaining in a direct manner 118 million jobs worldwide (3.8%), while over 313 million were considered as sustained indirectly (10%) (UNWTO 2019; WTTC 2020a).

The recent crisis generated by the COVID-19 pandemic brings into discussion the robustness of the tourism industry, as well as its intrinsic capacity to bounce back, especially for economies where tourism represents a key driver of growth and employment (UNWTO 2020b). At the time of writing this chapter (September 2020), there are no comprehensive final data on the impact of the COVID-19 pandemic on international tourism arrivals. However, the latest estimates of the World Tourism Organization indicate a temporary fall of 58–78%, depending on the time period of the gradual opening of international borders and the lifting of travel restrictions. This drop will likely translate into an overall loss of 850 to 1.1 billion international tourists and a loss of approximately $1 trillion in export revenues from tourism. According to the World Travel & Tourism Council, the current crisis will induce a drop between 30 and 62% in tourism-based GDP and jobs, which will account globally for a loss between $2686 and $5543 billion in revenues and between 98.2 and 197.5 million jobs (WTTC 2020b). At the time of writing the present study, no forecast was available on the long-term impact of the pandemic.

Despite the current pessimistic prognosis, previous crises have confirmed the high resilience performance of the travel and tourism sector. Even if this sector was among the most (if not the most) affected industries, displaying a vulnerability to various types of shocks (political, economic, or pandemics) (Papatheodorou and Pappas 2016; Scherzer et al. 2019; Sheppard and Williams 2016), it managed to bounce back and recover in shorter periods than other sectors (Romão 2020). For example, it took only 6 months for the tourism sector to recover after the September 11 attacks, 5 months after the SARS crisis in 2003, and approximately 10 months after the 2008 economic crisis. Therefore, more than ever, the need to understand the close relationship between tourism and regional resilience calls for sustaining efforts from academia and regional policy-makers to thoroughly scrutinize the short- and long-term shock effects on tourism, both globally and regionally.

The present study aims to introduce and analyze the new concept of tourism-led resilience capacity as a new anchor point for analysis and policy. It uses a comprehensive investigation of regions in the European Union to provide evidence-based findings and policy recommendations.

8.2   Tourism, Regional Development, and Resilience

The interdependencies between tourism and development have been evaluated and confirmed by multiple perspectives during the last five decades. The multiplier effects in the receiving regions, the high dynamics and efficiency in creating new jobs, the opportunities in terms of sustainable development for the lagging regions have promoted tourism sector as a priority in the long-term strategies for many countries and regions over the world. More recently, however, studies integrating a resilience-based approach instead a development-based one started to question the overall positive role of tourism activities. Tourism can induce economic resilience under certain conditions, but it can also represent a vulnerability factor, accelerating and amplifying the impact of a shock. For this reason, in-depth studies which can lead to a better understanding of tourism-resilience independences are required. This type of studies can actively contribute to more effective resilience-based policies from the perspective of the EU's objectives regarding sustainable development and territorial convergence.

8.2.1   *Tourism and Regional Development*

Besides the positive dynamics and the contribution to GDP growth and employment, several other important features make tourism a sector of great interest for national and regional strategies, especially for the lagging regions.

8.2.1.1   Tourism and Growth

Tourism activities are strongly connected with other industries (mainly handicraft, construction, food and beverage industry, agriculture, and transportation), which contribute to its multiplier effect (Pascariu and Ibănescu 2018). According to a WTTC study from 2012, the multiplier effect of the tourism sector is higher than that of other sectors such as communications, financial services, or education (WTTC 2012). Consequently, tourism can be seen as a key driver for growth and development (leading to the well-known tourism-led growth hypothesis) (Balaguer and Cantavella-Jorda 2002). The multiplier effect depends on a wide variety of factors (business environment, international openness, local industry competitiveness, economic diversification, the existence of value chains), and therefore the specialization of a region on tourism activities raises issues regarding dependency risks (foreign markets, foreign capital) and the diminishing role of other industrial sectors (Pascariu and Ibănescu 2018; Romão and Nijkamp 2017, 2018). However, numerous studies have confirmed the relationship between tourism and economic growth, justifying the use of tourism as a leverage mechanism for long-term economic growth (Balaguer and Cantavella-Jorda 2002; Brida et al. 2016; Pablo-Romero and Molina 2013; Perles-Ribes et al. 2017).

8.2.1.2 Tourism and Convergence

Tourism manages to capitalize to a greater extent the low- and medium-skilled workforce and can be easily introduced in less capital-intensive and less innovative destinations (Jussila and Järviluoma 1998). Thus, tourism can be an attractive alternative for lagging regions focused, for instance, on agricultural activities or low-tech industries (Boujrouf et al. 1998; Ibanescu 2015). Being a form of direct export, tourism diversifies the opportunities of these regions on international markets and improves their export performance, contributes significantly to public budgets, can increase the attractiveness of tourism destinations for foreign direct investments, and stimulates the development of the SME sector and local entrepreneurship (Roudi et al. 2019; Sanford Jr and Dong 2000; UNCTAD 2020). As a result, the European Union considers tourism not only a strategic sector for sustaining economic growth and stimulating the competitiveness of the European economy but also an important driver of regional convergence. A series of studies has highlighted the ability of tourism to actively contribute to the reduction of development gaps between countries or regions due to its strong linkages with other sectors and actors of local economies (Dwyer et al. 2000; Khan et al. 2020; Pascariu and Ibănescu 2018).

8.2.1.3 Tourism and Community

Tourism is a labor-intensive industry, displaying one of the highest capacities to generate new jobs and to respond to the global objectives on women and youth employment participation (Jussila and Järviluoma 1998), thus generating social structural transformation in local communities and reducing poverty. All these aspects confirm the transformative role that tourism has in destinations and the ability to influence community well-being and its resilience (Brankov et al. 2019; Croes 2014). During the 2008 financial crisis, employment in the hospitality sector was less affected compared to other economic sectors (ILO 2013), a supplementary confirmation that tourism is generally more resilient than other economic sectors and can be a source of resilience performance for regional economies.

8.2.1.4 Tourism and Environment

A controversial topic regarding the relation between tourism and regional development is represented by the environmental effects of the tourism activities. While this does not represent the main topic of our chapter, it should be mentioned that adverse environmental effects are currently seen as one of the main issues generated by tourism activities. As of today, the discourse regarding tourism impacts on environment is split between two approaches. On the one hand, tourism contributes to global warming by increasing CO_2 emissions; it is developing on the basis of high consumption of resources and energy, it generates considerable amounts of carbon-based pollution, it has a high degree of spatial concentration and density, and it contributes to environmental degeneration (Romao et al. 2017). On the other hand, the growing interest of tourists in environmental values leads to investments in projects for biodiversity conservation, reducing pollution and developing eco-markets and eco-products in accordance with the principles and conditions of sustainable development (Backen et al. 2020; Brankov et al. 2019).

Tourism is acknowledged in the EU policies on regional development and convergence as a leverage mechanism for economic growth and sustainable development. The European Union receives approximately 40% of worldwide international tourist arrivals and 31% of earnings, being the world's leading tourist destination (UNWTO 2020a). With a growth rate higher than the real economy GDP growth (2.3% in 2019, compared to 1.4%) and a contribution of 9.5% to total GDP, respectively 11.2% to total employment, 6% of EU overall exports and 22% of services exports, tourism is considered one of the most important and dynamic economic sectors in EU (UNWTO 2020a). In fact, many European regions have included tourism-orientated policies in their development strategies due to their high capacity of stimulating the economy of destinations. Moreover, in some regions, tourism has been considered a sector of smart specialization, benefiting from specific financial support for innovation and development (Del Vecchio and Passiante 2017).

At the same time, the European tourism industry is characterized by a high structural fragility in SMEs representing over 95% of all tourism enterprises and facing significant deficits in quality management, in access to information technology, in access to finance, and in integration into networks and clusters (UNWTO 2020a). The current crisis generated by COVID-19 pandemic, while different from the economic crisis of 2008/2009, shares the same swift and devastating impact on the tourism sector. The first months of pandemic have accentuated the fragility of European tourism, and annihilated its capacity to perform to economic growth. The loss of jobs, the increase in the number of bankruptcies, the reduction of the purchasing power of the population, the restrictions on the freedom of movement at international level and within the domestic market, the reduction of capital accumulation and investments all these factors contributed to a general slowdown of tourism.

Despite the tremendous shockwave, tourism managed to keep its attractiveness due to its high resilience capacity, the same capacity displayed following the crisis of 2008/2009. During that crisis, the growth rate of international tourist arrivals in the EU decreased to −3.9% in 2009, but it rose rapidly to 6.6% in 2010, stabilizing for the next period at an average close to 4%. Foreign tourists' expenditures in EU reached 291 billion euros in 2012, exceeding the pre-crisis level (265 billion euros in 2008) and rose to 375 billion euros in 2016, an increase of 41% in only 8 years, against the background of a crisis that led to negative rates of economic growth at EU level both in 2009 and 2012 (−4.4%, respectively −0.5%). Therefore, tourism embodies an element of stability in regional economic dynamics, being often considered a priority in growth policies and an extremely attractive sector for business. This stability is due to a characteristic which gained popularity during the last decades, namely the concept of resilience.

8.2.2   Tourism and Resilience: Theoretical Approaches and Empirical Evidence

Besides the positive effects of tourism on economic growth (Antonakakis et al. 2015; Balaguer and Cantavella-Jorda 2002; Brida et al. 2016; Schubert and Brida 2011), recent studies have suggested that due to its dynamism and strong connections with related economic branches, tourism activities manage to contribute to the increase in resilience capacity of the affected territories, an aspect which incited academics in asserting that tourism destinations should display resilience rather than growth (Cheer et al. 2019).

In regional science, resilience is commonly defined as the capacity of a system (city, region, country) to resist, absorb, and recover from a shock or a disturbance, bouncing back (returning to the pre-shock position) or bouncing forward, by a structural transformation towards a new growth pattern (Béné et al. 2014; Muštra et al. 2017; Reggiani et al. 2002). A similar perspective is supported by Martin and Sunley (2015) who offer a more exhaustive definition: the resilience is "the capacity [. . .] to withstand or recover from [. . .] shock to its developmental growth path, if necessary by undergoing adaptive changes to its economic structures and its social and institutional arrangements, so as to maintain or restore its previous developmental path, or transit to a new sustainable path characterized by a fuller and more productive use of its physical, human and environmental resources" (Martin and Sunley 2015). Therefore, resilience is seen as a process in an evolutionary approach, from resistance and absorption of a shock (*absorptive capacity*) to recovering and transformation (*adaptive capacity*). The reaction of a system is highly dependent on its vulnerabilities, its robustness (which will determine its absorptive capacity), respectively the transformational responses which will dictate the adjustment of the system to the shock while maintaining its main functions ("self-restoring equilibrium dynamics," "path dependency," "adaptation"). At the same time, the system could evolve and develop, based on a learning process, new structures (social or economic) and new functions more capable and more reactive to future shocks, thus enhancing a long-term territorial development potential ("positive adaptability," "prosilience," "evolutionary resilience," "adaptability") (Béné et al. 2014; Boschma 2015; Christopherson et al. 2010; Martin and Sunley 2015; Simmie and Martin 2010). In fact, this could be resumed as the distinction between short-term (absorption shocks) and long-term (adaptation vs. adaptability to shocks) approaches.

From an evolutionary perspective which refers to the capacity of the system to adopt new models of development, the resilience depends essentially on the institutional quality (Ezcurra and Rios 2019), industrial structures and linkages (Boschma

2015), innovation performance (Bristow and Healy 2018), financial arrangements (Belke et al. 2016), labor market structure (Stanickova and Melecký 2018), territorial capital (Fratesi and Perucca 2018), social capital, and local community (Mulligan et al. 2016). All the above-mentioned factors are also susceptible to core-periphery differentiations, which suggests a strong spatial component in resilience capacity, highly relevant for convergence policies undertaken by the EU (Pascariu and Tiganasu 2014).

The first mentions of the resilience concept in relation to tourism activities appeared in the '1990s (O'Hare and Barrett 1994), mostly in relation with environmental and economic shocks, or risk management. More and more studies recognized its importance, especially for policy-makers: "The management of unforeseeable and unpredictable situations is one of the 'strategic issues,' which lies within the responsibility of the top management of a (tourist) destination" (Innerhofer et al. 2018).

The current literature even sees regional resilience as highly connected with the notion of regional sustainability, and therefore, the attention given by scholars to the factors susceptible of positively inducing a resilience capacity is constantly expanding (Cellini and Cuccia 2015; Cheer et al. 2019). The importance of tourism resilience is even more prominent when we consider that people working in the tourism industry are more vulnerable to losing their jobs during shocks than people working in other sectors (Scherzer et al. 2019; Sheppard and Williams 2016). Recent studies found evidence of tourism-induced resilience capacity (Cellini and Cuccia 2015; Innerhofer et al. 2018), this positive impact of tourism upon regional resilience capacity being most likely induced by the high resilience of the tourism industry itself (Cellini and Cuccia 2015). Usually, tourism activities have been identified as very sensitive to the onset of an economic, natural, environmental, social, or military crisis; however, the tourism industry's after-shock recovery rate is higher than the recovery rate of most traditional economic sectors (agriculture, industry, or commerce) (Cellini and Cuccia 2015). Consequently, tourism is usually seen as a fail-safety mechanism for economic growth after a natural or socially induced shock.

The economic crisis from the late 2000s and the current crisis generated by the COVID-19 pandemic have revealed new challenges for tourism destinations such as the adequate management of socioeconomic risks and the reduction of their negative impacts on tourism flows and on regional development. These challenges questioned the role of tourism as driver for regional development and sustainability in destinations. Therefore, the academic discourse has shifted towards the concept of resilient tourism destination. Unfortunately, most of the recent studies on the topic have limited their findings to local levels, without going further to a regional or national analysis.

Thus, some key questions emerge in substantiating recovery policies, especially from the perspective of the challenges for the EU related to structural core-periphery differences, respectively: How resilient is the European tourism when put in a regional framework and how salient are core-periphery differences in terms of shock vulnerability and resilience performance? What are the characteristics of the regions with high tourism performance? How does tourism contribute to regional resilience and what are the implications for the core-periphery model?

Therefore, considering the connections between tourism industry and the destinations, as well as the drivers acting on tourism sector performance, our study aims to fill some gaps in the understanding of tourism resilience as well as tourism-generated effects on European economies and how tourism can be capitalized in European regional policies in order to stimulate resilience capacity. Consequently, the remaining part of this chapter addresses three complementary objectives to be empirically investigated:

1. O1: assessing and mapping the tourism resilience at regional (NUTS 2) level;
2. O2: testing the correlation between tourism resilience and regional characteristics previously identified in the literature as drivers of resilience;
3. O3: identifying the relation between tourism resilience and regional economic resilience.

For this study we have considered the fact that tourism and regional development are closely interrelated in EU policies, given the importance of tourism for the European economy and its contribution to the achievement of the EU's goals related to economic growth and job creation.

8.3   Methodological Approach

For the present study we took into consideration the EU regions (NUTS 2 statistical level). The data selected for analysis was the most recent available in terms of tourism indicators, economic resilience, innovation, competitiveness, and economic diversification. For tourism-based indicators, our study has selected as main variables tourism arrivals, overnight stays, number of establishments, and number of bed-places, while for describing the economic performance we choose mainly the behavior of EU regions during and after the 2008 economic crisis. Based on the availability of the data, it was decided to apply cross-section analysis and multiple linear regressions to evaluate the level of influence of tourism resilience on regional resilience capacity within the European Union. The sources for the data were represented by official OECD and EUROSTAT databases. It should be noted that the authors considered as number of arrivals in this study the sum of both domestic and international arrivals. This approach was adopted due to its exhaustive characteristics and because it is not dependent on the variations recorded on regional tourism markets.

In terms of shocks, the chapter is addressing the tourism-based responses in regional economies following the financial crisis of 2008–2012. We focused our research on two complementary axes:

1. Assessing the resilience of the tourism sector;
2. Estimating the role of tourism in the overall resilience performance of EU regions.

Main features of tourism and its regional policy implications in the European Union were also addressed. In terms of resilience indicators, we calculated two different indicators of tourism resilience:

1. Speedy tourism resilience (measuring the rapidity of a region in bouncing back to the same achievement levels as before the shock).
2. Sustained tourism resilience (measuring the capacity of a region of sustaining a positive growth for more years in the post-shock period);

The approach allowed the identification of the impact of tourism on regional resilience, but also the resilience of the tourism sector when facing external shocks. Data analysis of a statistical modeling nature was performed using IBM SPSS Statistics 21, while the cartographic part was realized with ESRI ArcMap 10 software.

8.4   The Intricate Manifestation of Tourism Resilience

According to the literature, and sustained by the statistical data, the tourism is impacted differently according to the shock typology (economic, environmental, political, pandemic) and their specific manifestations. In a similar manner, the process of tourism recovery is distinct at regional level, offering a complex picture in which the speed of recovery does not necessarily correlate with its stability over time, while the resilience of the tourism sector impacts distinctively the regional economic resilience.

8.4.1   Tourism Resistance: First Answer to Crisis: The Specific Case of the 2008/2009 Economic Crisis

The resistance designates the first reaction that a system manifests following a shock of any nature. Regarding the economic crisis, the tourism sector was one of the last sectors to feel the effects of global recession, the decline in international arrivals starting only in the second semester of 2008 (Papatheodorou et al. 2010; Smeral 2009), therefore displaying a slightly higher resistance than other sectors—like banking, for example. However, the shockwave expanded throughout 2009 as well, making 2009 the year with the lowest decrease globally in tourism arrivals by that time.

While the economic crisis affected the whole planet, its impact upon tourism arrivals manifested itself differently from region to region. Europe, Americas, and

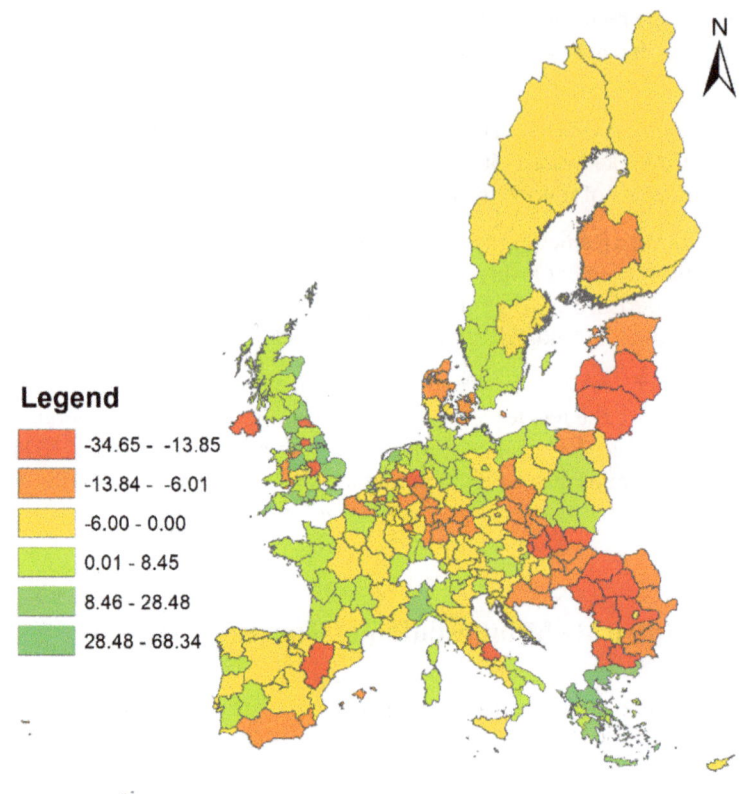

Fig. 8.1 Dynamics of tourist arrivals in 2009 in the European Union (Tourism resistance)

the Middle East were the most affected with −5.7%, −4.8%, respectively −4.9%, translated in a net loss of tens of millions of tourists (UNWTO 2010). Economically, the tourist loss altered considerably the European and American travel markets, which recorded net losses of −7%, respectively −10%. It was found that the instability and the job losses have affected seriously the travel decisions as well; therefore, a snowball effect had manifested throughout the travel sector (Alegre et al. 2013; Papatheodorou and Pappas 2016).

The European tourism sector, which is historically the largest and most mature, has been hit appreciably by the economic crisis. While a general shock wave can be observed throughout the European Union (which concentrates the vast majority of European arrivals) in 2009, the effect is not evenly distributed at a subregional level (Fig. 8.1). Central and Eastern destinations have been affected more severely than Western destinations, for example. Moreover, peripheral regions, especially from the South and the East, were more severely affected than the core regions, a feature which can be explained by the overall low capacity in responding to external shocks. For the Spanish regions, the intensity of the shock may be explained by the high reliance on the UK source market and the temporary drop of UK pound value

(UNWTO 2010). Greece represents a curious exception, however; it should be mentioned that most of Greece's NUTS 2 regions felt the shock wave 1 year later (2010). While the number of international tourists dropped by 6.4 in 2009 (UNWTO 2010), the apparently healthy values of internal arrivals maintained Greece, at least for 2009, in a positive dynamics.

At NUTS 2 level, only 106 regions displayed a medium or strong resistance (>0), a clear indication of the higher vulnerability of tourism towards the economic crisis. The European regions reacted heterogeneously during the first manifestations of the economic crisis, a variability due to socio-economic factors, as well as contrasted territorial capital. While during the last decade several tourist-centered studies tried to provide answers for different resistance values to the economic crisis, like the concept of "crisis-resistant tourist" (Hajibaba et al. 2015), which gained significant popularity and acknowledgment, the importance of the quality of life (Bronner and de Hoog 2014), or gender differences in tourism behavior (Ibanescu et al. 2018), these explanations are not sufficient for understanding the regional behavior and the discrepancies between territorial units.

Possibly the best tourist-centered explanation for regional variations was offered by Eugenio-Martin and Campos-Soria (2014) who mapped the probability of tourism expenditure cutback decision. The distribution of cutback decision, which in their approach depends on households' preferences for tourism under consumption changes and income variations (Eugenio-Martin and Campos-Soria 2014) shows consistent similarities with the map of tourism resistance (Fig. 8.1), with the highest probabilities of cutback decisions in Eastern Europe, especially in Romania, Bulgaria, and Hungary, as well as Southern Italy and Spain. However, additional territorial-based explanations, like the maturity of the tourism sector, the diversity of the tourism network, the existence or absence of immediate strategies, should be taken into consideration as well (Fratesi and Perucca 2018; Romao and Neuts 2017; Romão and Nijkamp 2018). Nonetheless, the resistance is illustrating merely the first answer of tourism industry to the shock; in order to understand the complexity of the relations between the tourism industry and the overall economic impact, as well as the mid-term dynamics, a closer look should be taken at the recovery pattern, namely at the resilience performance displayed by tourism destinations.

8.4.2 Speedy Tourism Resilience: The High Capacity of Tourism to Bounce Back

The indicator of speedy tourism resilience, which can be translated as the speed of recovery of the European regions in managing to reach the pre-shock values of tourism arrivals (the higher the indicator, the quicker the recovery), shows a powerful comeback of tourism arrivals all over EU.

Most of NUTS 2 regions managed to reach the pre-shock level of arrivals within 1–3 years. In fact, 183 regions display a high speed of recovery (within 1–2 years),

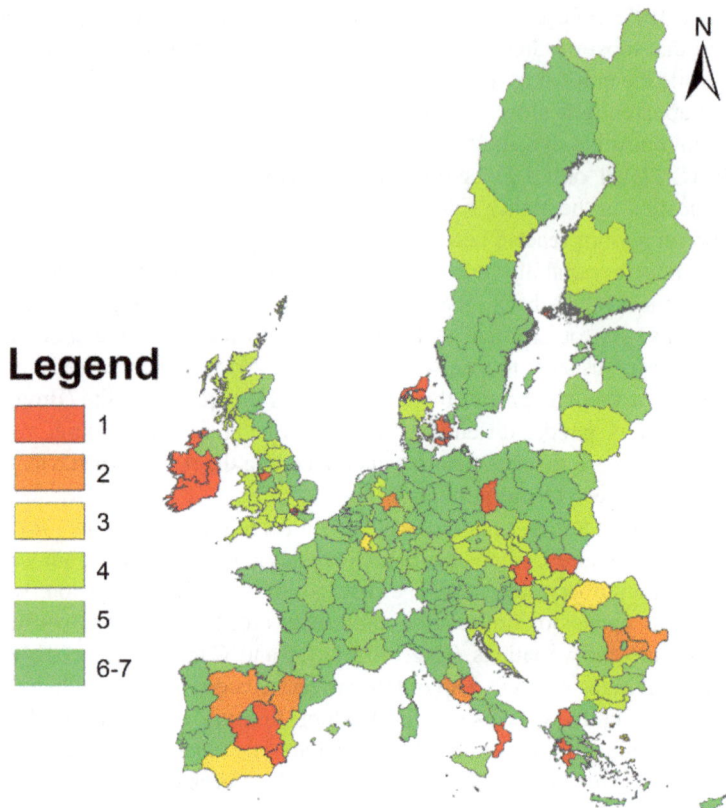

Fig. 8.2 Indicator of speedy tourism resilience (The value of the indicator is inversely proportional to the number of years necessary for the region to reach the pre-shock values in tourist arrivals; 7 = 1 year, 6 = 2 years, 5 = 3 years, 4 = 4 years, 3 = 5 years, 2 = 6 years, 1 = the regions did not reach the pre-shock values during the period of study)

and only 22 regions display a very low speed of recovery (over 6 years) (Fig. 8.2). The results support previous findings which claimed that the tourism sector has an overall quick recovery from the economic crisis (Cellini and Cuccia 2015) and encourage its application as a fail-safety mechanism for economic growth after a natural or socially induced-shock. Two observations should be made regarding the distribution of this indicator: first, while a low speed of recovery seems to appear only sporadically, countries like Romania, Ireland, and Spain display overall lower values and higher number of regions with a low speed of recovery; second, just like in the case of tourism resistance, a clear differentiation between core and peripheral regions can be noted, especially with regard to the national scale. While this indicator emphasizes the rapid recovery of tourism destinations, it does not account for the overall behavior.

8.4.3 *Sustained Tourism Resilience*

The second indicator of tourism resilience (sustained tourism resilience), which indicates the number of years with positive dynamics during the 6 years of post-shock period (2010–2015), displayed a more nuanced map (Fig. 8.3).

Only 113 regions scored a high value of sustained tourism resilience (5+ years of positive growth in the aftermath of the economic crisis), implying that sustained tourism resilience is harder to achieve than speedy tourism resilience. The significant differences between the two indicators of tourism resilience suggest that the attention of policy-makers should focus on both speedy recovery and sustained resilience in order to properly tackle the crisis manifestations in the tourism sector.

The map of sustained tourism resilience displays a more heterogeneous pattern for the European Union, with a mix of highly and medium resilient regions. Recent

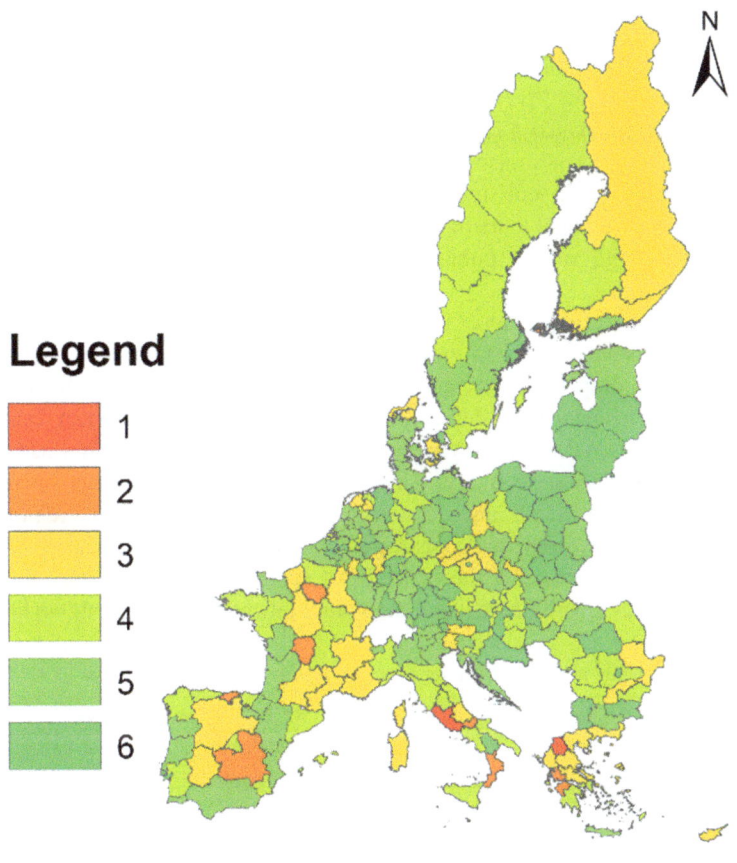

Fig. 8.3 3 Indicator of sustained tourism resilience (2010–2015) (The value of the indicator expresses the number of years recording positive tourism dynamics between 2010 and 2015)

studies looking into the regional behavior of European tourism during the last decades found higher growth rates for less-developed regions, while the most developed regions were more resilient (Romão 2020).

8.4.4 Regional Features and Tourism Resilience

In order to have an in-depth assessment of the relation between tourism and regional resilience, we proceeded to a more complex analysis based on specific tourism indicators (length of stay), regional performance indicators (regional competitiveness, regional innovation, regional diversification), and the time of recovery. The choice of the indicators was motivated by previous studies underlying their role in moderation the relation between tourism activities and economic performance. The regional diversification and the regional competitiveness play an important role in amplifying the tourism multiplier effect (Pascariu and Ibănescu 2018), while the innovation is capable of smoothing the recovery of tourism at regional level (Del Vecchio and Passiante 2017).

The correlation matrix (Table 8.1) confirms the supporting role of tourism in economic resilience and it also indicates a significant relation between the competitiveness of regions and tourism resilience. Economic diversification and innovation seem to increase regional tourism resilience to economic shocks as well, our finding being in line with results published by Luthe et al. (2012) and Romão and Nijkamp (2018).

Furthermore, the correlation matrix delineates two major findings: First, the indicator of sustained tourism resilience seems to be more connected with the diversification, competitiveness, and innovation capacity of a region than the indicator of speedy resilience. Most likely, the indicator of sustained tourism resilience is highly dependent on the above-mentioned regional features, while also managing to contribute to a quicker recovery. This could also suggest that the interrelation between tourism resilience and economic resilience is stronger in regions with higher economic diversity, and consequently, a higher multiplier effect (Pascariu and Ibănescu 2018). Second, the indicator of speedy tourism resilience displays a stronger correlation with the regional innovation, therefore supporting the huge role

Table 8.1 Correlation matrix between resilience indicators and regional features

	B/E	LOS	RCI	HHI	RII	YCR
Indicator of sustained tourism resilience	.005	-.071	.219**	.144*	.195**	-.349**
Indicator of speedy tourism resilience	.089	-.060	.129*	-.106	.330**	-.090

LOS length of stay, *RCI* regional competitiveness index, *RII* regional innovation index, *HHI* Herfindahl-Hirschman index (HHI) = A common measure of market concentration, *YCR* year crisis recovery = number of years for the regions to recover after crisis
**Correlation significant at the 0.01 level (2-tailed)

of innovation in contributing to the tourism restart following a shock (Del Vecchio and Passiante 2017), probably through innovative products, rapid adaptation to the societal needs, and quicker access to innovative instruments. However, while the innovation helps a quicker tourism recovery, the diversification and competitiveness of a region are extremely important for securing a long-term recovery following a shock.

Subsequently, indicators such as the levels of tourism establishments or the length of stay seem to have little impact on tourism resilience, thus supporting the assertion that the regional structure plays a more salient role than the tourism structure in providing tourism resilience. Our preliminary results seem to be supporting the principle of integrating tourism into growth and resilience-oriented policies, especially in relation with strategies of economic diversification, competitiveness and innovation increase.

8.5 Discussions, Conclusions, and Implications for Policies

Our results confirm the existence of a strong connection between tourism and economic development, beyond the classical tourism-led growth hypothesis. While previous studies focused mainly on the relation between tourism and economic growth (Antonakakis et al. 2015; Balaguer and Cantavella-Jorda 2002; Brida et al. 2016; Perles-Ribes et al. 2017), our study introduces and confirms the tourism-led resilience hypothesis. The first analyses of the data back up the assumption that tourism activities have the potential to enhance the resilience capacity at regional level within the European Union. Tourism can be seen as a fail-safety mechanism for economic recovery after a major shock.

What could be the implications for regional policy-makers of a tourism-led resilience strategy and how can tourism activities be used for easing the answer to current or future disruptive shocks? The tourism industry as well as whole regional economic networks are in acute need of knowledge for informed decision-making and for innovative strategies in order to effectively respond to worldwide disruptions. Based on previous shocks which generated major crisis, the tourism sector is expected to recover quicker and at a relatively higher speed than other economic branches. Moreover, certain forms of travel are expected to perform better in the post-shock recovery. Domestic tourism is expected to recover faster than international tourism, while travelling to visit friends and relatives tends to perform better in the recovery period than more specialized forms of tourism, like business travel.

Our study offers three major suggestions for regional policy-makers in order to support the building of recovery strategies. Firstly, sustained resilience represents a far better goal than fast resilience for tourism destinations following a powerful shock. Given the apparent inevitability of future disruptions—either economic or pandemic—, the attention of tourism planning actors and scholars should not be seized by whether the sector will be affected or not, since its immunity is illusory, but rather by how the sector will manage to adapt and recover in a sustainable manner afterwards, with a particular focus on building resilient industries and destination capabilities. Recent papers looking into the COVID-19 influence on tourism (Hall et al. 2020; Nepal 2020) are addressing the same question regarding the post-pandemic tourism micro-cosmos: how sustainable will the recovery be? Moreover, in a very recent review of research on tourism risk, crisis, and disaster management (Ritchie and Jiang 2019), the authors specifically call for future research on a better understanding of sustainable tourism resilience and the factors that influence it.

Secondly, the study confirms the powerful effect of innovation on tourism resilience. As stated in recent studies looking into the role of innovation and creativity on recovery speed after the economic crisis (Mazilu et al. 2020), policy-makers should focus more on enhancing the innovation score of their territory, as this could prove to be extremely useful for speeding up the process and building a long-lasting recovery.

Thirdly, the diversification of regional economy provides additional solutions following a disruption. Due to the close ties between tourism activities and other economic sectors, this could significantly speed up the recovery wave throughout the economy (Ibanescu et al. 2020). The economic diversification works as a solid engine for the reduction of the core-periphery hiatus. Regional policy makers should develop targeted interventions aimed at enhancing the resilience capacity in periph-eral and lagging regions in order to reduce regional disparities and increase tourism competitiveness throughout the European Union.

However, more research on this topic needs to be undertaken before the associ-ation between tourism activities and regional resilience capacity is clearly under-stood. Additional studies should also focus on more internal factors, such as the importance of trust, social capital, internal leadership capacity, and attitudinal commitment.

Acknowledgments This work was supported by a grant of Romanian Ministry of Research and Innovation, CNCS—UEFISCDI, project number PN-III-P4-ID-PCCF-2016-0166, within PNCDI III project ReGrowEU—Advancing ground-breaking research in regional growth and development theories, through a resilience approach: towards a convergent, balanced, and sustainable European Union.

References

Alegre J, Mateo S, Pou L (2013) Tourism participation and expenditure by Spanish households: the effects of the economic crisis and unemployment. Tour Manag 39:37–49

Antonakakis N, Dragouni M, Filis G (2015) How strong is the linkage between tourism and economic growth in Europe? Econ Model 44:142–155. https://doi.org/10.1016/j.econmod.2014.10.018

Backen S, Whittlesea E, Loehr J, Scott D (2020) Tourism and climate change. J Sustain Tour 28 (10):1603–1624. https://doi.org/10.1080/09669582.2020.1745217

Balaguer J, Cantavella-Jorda M (2002) Tourism as a long-run economic growth factor: the Spanish case. Appl Econ 34(7):877–884

Belke AH, Haskamp U, Setzer R (2016) Bank efficiency and regional growth in Europe: new evidence from micro-data. ECB Working Paper No.1983. https://doi.org/10.2866/84555

Béné C, Newsham A, Davies M, Ulrichs M, Godfrey-Wood R (2014) Review article: resilience, poverty and development. J Int Dev 26(5):598–623. https://doi.org/10.1002/jid.2992

Boschma R (2015) Towards an evolutionary perspective on regional resilience. Reg Stud 49 (5):733–751

Boujrouf S, Bruston M, Duhamel P, Knafou R, Sacareau I (1998) Les conditions de la mise en tourisme de la haute montagne et ses effets sur le territoire. L'apport d'une comparaison entre le Haut-Atlas et le Népal mise en perspective à l'aide du précédent alpin (exemple du massif du Mont-Blanc). Revue de Géographie Alpine 86(1):67–82

Brankov J, Penjišević IB, Ćurčić N, Živanović B (2019) Tourism as a factor of regional development: community perceptions and potential Bank support in the Kopaonik National Park (Serbia). Sustainability 11(22):6507

Brida JG, Cortes-Jimenez I, Pulina M (2016) Has the tourism-led growth hypothesis been validated? A literature review. Curr Issue Tour 19(5):394–430. https://doi.org/10.1080/13683500.2013.868414

Bristow G, Healy A (2018) Innovation and regional economic resilience: an exploratory analysis. Ann Reg Sci 60(2):265–284

Bronner F, de Hoog R (2014) Crisis resistance of tourist demand: the importance of quality of life. J Travel Res 55(2):190–204. https://doi.org/10.1177/0047287514541006

Cellini R, Cuccia T (2015) The economic resilience of tourism industry in Italy: what the 'great recession' data show. TMP 16:346–356. https://doi.org/10.1016/j.tmp.2015.09.007

Cheer JM, Milano C, Novelli M (2019) Tourism and community resilience in the Anthropocene: accentuating temporal overtourism. J Sustain Tour 27(4):554–572. https://doi.org/10.1080/09669582.2019.1578363

Christopherson S, Michie J, Tyler P (2010) Regional resilience: theoretical and empirical perspectives. Camb J Reg Econ Soc 3(1):3–10

Croes R (2014) The role of tourism in poverty reduction: an empirical assessment. Tour Econ 20 (2):207–226

Del Vecchio P, Passiante G (2017) Is tourism a driver for smart specialization? Evidence from Apulia, an Italian region with a tourism vocation. J Destin Mark Manag 6(3):163–165. https://doi.org/10.1016/j.jdmm.2016.09.005

Dwyer L, Forsyth P, Madden J, Spurr R (2000) Economic impacts of inbound tourism under different assumptions regarding the macroeconomy. Curr Issue Tour 3(4):325–363

Eugenio-Martin JL, Campos-Soria JA (2014) Economic crisis and tourism expenditure cutback decision. Ann Tour Res 44:53–73. https://doi.org/10.1016/j.annals.2013.08.013

Ezcurra R, Rios V (2019) Quality of government and regional resilience in the European Union. Evidence from the great recession. Pap Reg Sci 98(3):1267–1290

Fratesi U, Perucca G (2018) Territorial capital and the resilience of European regions. Ann Reg Sci 60(2):241–264

Hajibaba H, Gretzel U, Leisch F, Dolnicar S (2015) Crisis-resistant tourists. Ann Tour Res 53:46–60. https://doi.org/10.1016/j.annals.2015.04.001

Hall CM, Scott D, Gössling S (2020) Pandemics, transformations and tourism: be careful what you wish for. Tour Geogr 22(3):577–598. https://doi.org/10.1080/14616688.2020.1759131

Ibanescu B (2015) Consequences of peripheral features on tourists' motivation. The case of rural destinations in Moldavia, Romania. J Settlements Spat Plann 4:191–197

Ibanescu B-C, Stoleriu OM, Gheorghiu A (2018) Gender differences in tourism behaviour in the European Union. East J Eur Stud 9(1):23

Ibanescu B-C, Eva M, Gheorghiu A (2020) Questioning the role of tourism as an engine for resilience: the role of accessibility and economic performance. Sustainability 12(14):5527

ILO (2013) Economic crisis, international tourism decline and its impact on the poor. https://www.ilo.org/wcmsp5/groups/public/%2D%2D-ed_dialogue/%2D%2D-sector/documents/publication/wcms_214576.pdf

Innerhofer E, Fontanari M, Pechlaner H (2018) Destination resilience: challenges and opportunities for destination management and governance. Taylor & Francis. https://books.google.ro/books? id=6FFHDwAAQBAJ

Jussila H, Järviluoma J (1998) Extracting local resources: the tourism route to development in Kolari, Lapland, Finland. In: Neil C, Tykkyläinen M (eds) Local economic development: a geographical comparison of rural community restructuring. UN University Press, Tokyo, pp 269–289

Khan A, Bibi S, Lorenzo A, Lyu J, Babar ZU (2020) Tourism and development in developing economies: a policy implication perspective. Sustainability 12(4):1618

Luthe T, Wyss R, Schuckert M (2012) Network governance and regional resilience to climate change: empirical evidence from mountain tourism communities in the Swiss Gotthard region. Reg Environ Chang 12(4):839–854. https://doi.org/10.1007/s10113-012-0294-5

Martin R, Sunley P (2015) On the notion of regional economic resilience: conceptualization and explanation. J Econ Geogr 15(1):1–42

Mazilu S, Incaltarau C, Kourtit K (2020) The creative economy through the lens of urban resilience. An analysis of Romanian cities. Transylv Rev Adm Sci 16(59):77–103

Mulligan M, Steele W, Rickards L, Fünfgeld H (2016) Keywords in planning: what do we mean by 'community resilience'? Int Plan Stud 21(4):348–361

Muštra V, Šimundi B, Kuliš Z (2017) Effects of smart specialization on regional economic resilience in EU. Rev Estud Reg 110:175–195

Nepal SK (2020) Travel and tourism after COVID-19 – business as usual or opportunity to reset? Tour Geogr 22(3):646–650. https://doi.org/10.1080/14616688.2020.1760926

O'Hare G, Barrett H (1994) Effects of market fluctuations on the Sri Lankan tourist industry: resilience and change, 1981–1991. Tijdschr Econ Soc Geogr 85(1):39–52. https://doi.org/10.1111/j.1467-9663.1994.tb00672.x

Pablo-Romero MDP, Molina JA (2013) Tourism and economic growth: a review of empirical literature. Tour Manag Perspect 8:28–41

Papatheodorou A, Pappas N (2016) Economic recession, job vulnerability, and tourism decision making: a qualitative comparative analysis. J Travel Res 56(5):663–677. https://doi.org/10.1177/0047287516651334

Papatheodorou A, Rosselló J, Xiao H (2010) Global economic crisis and tourism: consequences and perspectives. J Travel Res 49(1):39–45

Pascariu GC, Ibănescu B-C (2018) Determinants and implications of the tourism multiplier effect in EU economies. Towards a Core-periphery pattern? Amfiteatru Econ 20(12):982–997

Pascariu GC, Tiganasu R (2014) Tourism and sustainable regional development in Romania and France: an approach from the perspective of new economic geography. Amfiteatru Econ J 16 (Special No. 8):1089–1109

Perles-Ribes JF, Ramón-Rodríguez AB, Rubia A, Moreno-Izquierdo L (2017) Is the tourism-led growth hypothesis valid after the global economic and financial crisis? The case of Spain 1957–2014. Tour Manag 61:96–109

Reggiani A, De Graaff T, Nijkamp P (2002) Resilience: an evolutionary approach to spatial economic systems. Netw Spat Econ 2(2):211–229. https://doi.org/10.1023/A:1015377515690

Ritchie BW, Jiang Y (2019) A review of research on tourism risk, crisis and disaster management: launching the annals of tourism research curated collection on tourism risk, crisis and disaster management. Ann Tour Res 79:102812. https://doi.org/10.1016/j.annals.2019.102812

Romão J (2020) Tourism, smart specialisation, growth, and resilience. Ann Tour Res 84:102995. https://doi.org/10.1016/j.annals.2020.102995

Romao J, Neuts B (2017) Territorial capital, smart tourism specialization and sustainable regional development: experiences from Europe. Habitat Int 68:64–74

Romão J, Nijkamp P (2017) Spatial-economic impacts of tourism on regional development: challenges for Europe. No. 2017_01. University of Evora CEFAGE-UE (Portugal)

Romão J, Nijkamp P (2018) Spatial impacts assessment of tourism and territorial capital: a modelling study on regional development in Europe. Int J Tour Res 20(6):819–829

Romao J, Kourtit K, Nijkamp P (2017) The Smart City as a common place for tourists and residents. Cities 78:67–75. https://doi.org/10.1016/j.cities.2017.11.007

Roudi S, Arasli H, Akadiri SS (2019) New insights into an old issue–examining the influence of tourism on economic growth: evidence from selected small island developing states. Curr Issue Tour 22(11):1280–1300

Sanford DM Jr, Dong H (2000) Investment in familiar territory: tourism and new foreign direct investment. Tour Econ 6(3):205–219

Scherzer S, Lujala P, Ketil J (2019) International journal of disaster risk reduction a community resilience index for Norway : an adaptation of the baseline resilience indicators for communities (BRIC). Int J Disaster Risk Reduct 36:101107. https://doi.org/10.1016/j.ijdrr.2019.101107

Schubert SF, Brida JG (2011) Dynamic model of economic growth in a small tourism driven economy. In: Matias A, Nijkamp P, Sarmento M (eds) Tourism economics. Springer, New York, pp 149–168

Sheppard VA, Williams PW (2016) Journal of hospitality and tourism management factors that strengthen tourism resort resilience. J Hosp Tour Manag 28:20–30. https://doi.org/10.1016/j.jhtm.2016.04.006

Simmie J, Martin R (2010) The economic resilience of regions: towards an evolutionary approach. Camb J Reg Econ Soc 3(1):27–43

Smeral E (2009) The impact of the financial and economic crisis on European tourism. J Travel Res 48(1):3–13

Stanickova M, Melecký L (2018) Understanding of resilience in the context of regional development using composite index approach: the case of European Union NUTS-2 regions. Reg Stud Reg Sci 5(1):231–254

UNCTAD (2020) COVID-19 and tourism assessing the economic consequences. https://unctad.org/en/PublicationsLibrary/ditcinf2020d3_en.pdf

UNWTO (2010) International tourism highlights, 2010 edition. https://www.e-unwto.org/doi/epdf/10.18111/9789284413720

UNWTO (2019) International tourism highlights, 2019 edition. https://www.e-unwto.org/doi/book/10.18111/9789284421152

UNWTO (2020a) European Union tourism trends. https://www.e-unwto.org/doi/epdf/10.18111/9789284419470

UNWTO (2020b) International tourism and covid-19. https://www.unwto.org/international-tourism-and-covid-19

WTTC (2012) The comparative economic impact of tourism 2012. https://www.wttc.org/economic-impact/benchmark-reports/the-comparative-economic-impact-of-travel-tourism/

WTTC (2020a) Economic impact reports. https://wttc.org/Research/Economic-Impact

WTTC (2020b) Recovery scenarios 2020 & economic impact from COVID-19. https://wttc.org/Research/Economic-Impact/Recovery-Scenarios-2020-Economic-Impact-from-COVID-19

Chapter 9
Cross-Border Sustainable Tourism Development for Busan-Fukuoka Megapolitan Cluster in Northeast Asia

Jaewon Lim, Yasuhide Okuyama, and DooHwan Won

Abstract Using the historic annual visitor data, this study analyzed the recent trends of inbound visitors to Busan and Fukuoka Metros and estimated the economic impact of Japanese (or Korean) visitor spending in Busan (or Fukuoka) Metro. Trend analysis clearly indicates the rapid growth in the number of Korean visitors to Fukuoka since 2011, whereas the growth in the number of Japanese visitors is sluggish and even on a decline since the early 2010s. With the recent surge of Chinese visitors to Japan and Korea, the relative shares have declined, but these two neighboring countries still exchange a large number of visitors. Busan and Fukuoka Metro governments have been working together for the successful launch of Fukuoka-Busan Supra-Regional Economic Zone; however, they have not yet shown any tangible results, especially in the tourism industry. The economic impact analyses in this study show that the limited inter-industrial linkages of the tourism industry between Yeongnamkwon and Kyushu, the two closest cross-border neighboring regions between Korea and Japan. Two cross-border metros (Busan and Fukuoka) in these regions should initiate a set of strategies for tourism destination development collaboratively. By codeveloping and marketing various tourism resources such as islands between two metros, Busan-Fukuoka megapolitan cluster can be recognized as the Mediterranean in Northeast Asia and grows as a destination for the growing demand for cruise tourism in Asia, especially among Chinese tourists. Multimodal travel packages for domestic travelers can be extended to include Busan for Japanese travelers and Fukuoka for Korean travelers. Two

J. Lim (✉)
School of Public Policy and Leadership, Greenspun College of Urban Affairs, University of Nevada, Las Vegas, NV, USA
e-mail: jaewon.lim@unlv.edu

Y. Okuyama
Graduate School of Social System Studies, University of Kitakyushu, Kitakyushu, Japan
e-mail: okuyama@kitakyu-u.ac.jp

D. Won
Department of Economics, College of Economics and International Trade, Pusan National University, Busan, South Korea
e-mail: doohwan@pusan.ac.kr

© The Author(s), under exclusive licence to Springer Nature Singapore Pte Ltd. 2021
S. Suzuki et al. (eds.), *Tourism and Regional Science*, New Frontiers in Regional Science: Asian Perspectives 53, https://doi.org/10.1007/978-981-16-3623-3_9

metro governments are required to build a comprehensive understanding of the travelers' behavior in both destinations and to develop tourism products by effectively linking the two destinations. Busan-Fukuoka tourism bureau can be formed and it can develop a common platform and survey instruments to collect information from visitors.

Keywords Megapolitan cluster · Busan-Fukuoka · Adaptive tourism · Inbound visitor · Economic impact

9.1 Introduction

Japan and the Republic of Korea (ROK) are the geographically closest neighbors in Northeast Asia with the rapidly increasing visitations between the two countries. The two closest metropolitan areas are Fukuoka in Japan and Busan in Korea and the distance between the two is 214 km across the ocean. It takes only 45 min by flight or 3 h by ferry, much closer than the distance between Busan and Seoul and between Fukuoka and Osaka. Osaka is the closest MMA (Major Metropolitan Area) to Fukuoka among the more populous MMAs than Fukuoka and located 610 km away. Seoul is the only metropolitan area bigger than Busan and located 325 km away. Though Busan and Fukuoka are geographically close neighboring metros, the interaction between the two metros has not been fully utilized since they are located across the national sea border.

Given the enhanced interregional cohesion, these two metropolitan areas across the national border will be able to promote regional economic integration effectively. The effectively integrated regional economy will further promote trade of goods and services and develop the integrated markets for goods and services. Nelson and Lang (2011) defined the 10 megapolitan clusters that encompass the 23 megapolitan areas in the contiguous 48 states of the United States. They found that the enhanced interregional cohesion among multiple megapolitan areas is the key to form and promote the sustainable growth of a megapolitan cluster. For instance, the fastest-growing *Southwest* megapolitan cluster in the U.S. (linking the three megapolitan areas, namely Southern California, Las Vegas, Sun Corridor across three states) has strengthened the regional cohesion through the expanded flows of people, goods, and services. Tourism within *Southwest* megapolitan cluster plays a crucial role in enhancing the interregional cohesion caused by the increasing flows of tourists and the integrated services of the tourism industry in the cluster. Another successful regional integration that promoted sustainable regional growth is *Øresund*, a cross-border region between Sweden and Denmark. *Øresund* region is composed of the Skåne county in Southern Sweden and Greater Copenhagen on the Danish side and has a population of 3.5 million. A stimulus for the integrated regional growth was the construction of the *Øresund* bridge that links Copenhagen (Danish side) and Malmö (Swedish side) of the strait between two countries. With the opening of the bridge in 2000, traffic had increased by 34% between the two cities (Garlick et al. 2006) that contributed to the regional integration between the two metros across the national border. Along with biotechnology, pharmaceuticals and health, and information technology and communications, the "*tourism, culture and recreation*" industry is one of the key industries in the *Øresund* region (Garlick et al. 2006).

Since 2011, inbound tourists to Japan started to grow rapidly thanks to the government policies to stimulate the tourism industry by making Japan more accessible for foreign visitors and to promote the tourism industry as a catalyst for revitalizing regional economies in less-population nonmetropolitan areas (Andonian et al. 2016). According to JNTO (Japan National Tourism Organization),[1] inbound tourism to Japan (foreign visitors to Japan) has grown from 6.2 million in 2011, to 19.7 million in 2015 and 31.2 million in 2018. The Japanese government set the goal of 40 million inbound visitors to Japan in 2020. If successful, the increased demand of foreign visitors is expected to stimulate sustainable regional economic growth in Japan. The total number of foreign visitors to Japan had grown by 284% between 2011 and 2017, while the matching growth rate of foreign visitors arriving in Fukuoka (inbound visitors through Fukuoka Airport and Hakata Port) for the same period was 304%. This is mainly driven by the rapid growth of Korean visitors to Fukuoka in recent years. The growth rate of Korean visitors arriving in Fukuoka is 412% for the 2011–2017 period. In other words, the tourism industry in Fukuoka prefecture has mainly benefitted from the rapidly growing number of Korean visitors. However, the number of Japanese visitors arriving in Busan (inbound visitors through Gimhae Airport and Port of Busan) had decreased by 20% for the same period (2011–2017). For the 2011–2017 period, the total number of foreign visitors to Korea had increased by 36%, while the number of foreign visitors to Busan had not grown with virtually no change. Still, the increasing Chinese visitors to Busan had compensated the 20% loss of Japanese visitors between 2011 and 2017. The tourism industry in Busan is mainly designed to attract domestic visitors from SMA (Seoul Metropolitan Area) with oceanic sports along its beaches and food and beverage services. Busan's tourism industry needs to develop a set of strategies to attract more visitors from Fukuoka-Kitakyushu Greater Metropolitan Region, home for 5.1 million people, which is 111 km closer than Seoul.

This chapter aims to identify the recent development trends of tourism industries in Busan and Fukuoka, estimate the economic impact of foreign visitor spending in Busan and Fukuoka, and discuss potential collaborative strategies to create tourism resources for Busan-Fukuoka megapolitan cluster across the strait between Korea and Japan.

In the following Sect. 9.2, the summary of tourism literature and regional development are discussed. Section 9.3 introduces the methodologies to identify the recent trend of tourism industries in Busan and Fukuoka and to estimate the economic impact of visitor spending. The following Sect. 9.4 shares the identified trends and estimation results. Finally, Sect. 9.5 wraps up the chapter with policy

---

[1]Annual inbound visitor statistics is available from the following link, https://www.tourism.jp/en/tourism-database/stats/inbound/

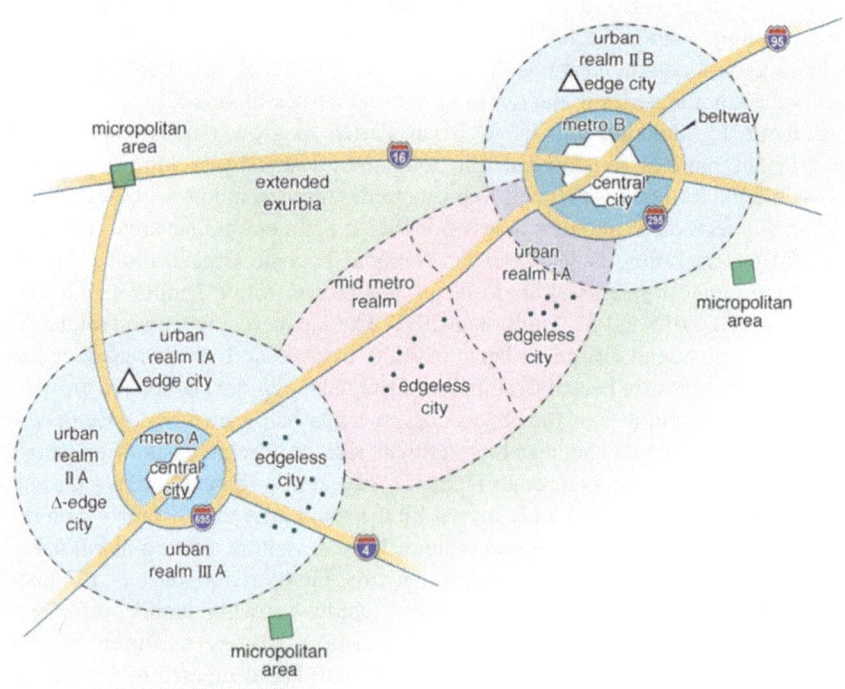

Fig. 9.1 Anatomy of New Metropolis (Lang and Knox 2009)

implications for the collaborative tourism industrial development for Busan-Fukuoka megapolitan cluster.

9.2 Literature Review

During the transition period to the post-industrial economic structure since the early 1980s, there has been the flurry of terminology in urban forms with variable geometry. For instance, *"LA School"* focused on disjointness, disorder, and variability of restructuring outcomes in the 2000s. Other urban scholars developed evolving ideas on urban/metropolitan forms: *new urban development at "Inter-Metropolitan Periphery"* (Berry 1980), *"Galactic Metropolis" with disjoint and decentralized urban forms* (Berry 1980), *"Technoburb" with no clear distinctions among zones* (Fishman 1987), *"100-mile city" with emerging metropolitan fringes* (Sudjic and Sayer 1992), *6 common types of nodes within the polycentric metropolitan form* (Hall and Pain 2006), and *stasis, rather than volatility, as a common pattern of urban forms* (Hackworth 2005).

Among others, Lang and Knox (2009) described the *New Metropolis* anatomy with emerging clusters of decentralized employment centers as shown in Fig. 9.1.

Residential fabric is based on the "Quasi-urbanized suburbs" (Frey 2003) and "Boomburbs" (Lang and LeFurgy 2007). While the former is residential areas with the concentrations of poverty and a growing share of single-person households and seniors, the latter are fast-growing suburbs with a population size of over 100,000.

In the proposed anatomy, Lang and Knox (2009) expand the geographic scope to bind multiple megalopolitan areas as part of a megalopolitan region linked by interstate highways, beltways, arterial highways, and urban freeways. The proposed megalopolitan region is composed of multiple metropolitan areas, principal cities, and micropolitan areas. Examples of colliding metropolitan areas include Dallas-Fort Worth in the 1960s, Washington-Baltimore in the1980s, and more recently during the 2010s, Phoenix-Tucson, Tampa-Orlando, and San Antonio-Austin. Later, Nelson and Lang (2011) developed the concept of megapolitan cluster linking multiple megapolitan areas and suggested 10 megapolitan clusters in the U.S. The clusters extended into 37 states occupying roughly 20% of land in the lower 48 states, while they accommodated about 65% of total U.S. population in 2010. The share of the U.S. population in the clusters is expected to grow to 70% by 2040. It is estimated to have 29.3 million more residents between 2010 and 2040. The fore-casted growth will consolidate the links among principal cities and micropolitan areas. A recent study by Lang et al. (2020) also explores the evolving concept of the megapolitan clusters.

The cohesive factors within megapolitan clusters include trade (goods movement), business linkages (flows of services), migration (flows of the population), tourists' visitation, cultural commonality, and physical environment. The main cohesive factors of megapolitan clusters differ from the traditional daily commuting sheds. The geographic scope of megapolitan clusters is too big for daily commuting. However, the share of commuters between metropolitan areas within megapolitan clusters increases, but it is impossible to detect how often commuters make trips per week. Tourism industries of metropolitan areas that attract tourists from other metropolitan areas within a megapolitan cluster emerge as an important factor to enhance regional cohesion of a megapolitan cluster. One good example is the Southwest megapolitan cluster in the U.S. that links Southern California, Las Vegas, and Sun Corridor of Arizona as defined by Nelson and Lang (2011). Kwon et al. (2020) estimated the economic impact of Southern California visitors' spending in Las Vegas and identified at least 26% of the annual visitors to Las Vegas was from the 10 counties in Southern California in 2010 and 2011. Industrial linkages between Southern California and Las Vegas through entertainment and tourism industries have strengthened the regional cohesion between the two metros. Lang et al. (2020) also demonstrated how the evolving concept of the megapolitan area can be applied to the rapidly growing Southwest Triangle megapolitan cluster in the U.S in the future.

In the fourth phase of the promotion of strategic industries (2014-ongoing), the Busan Metropolitan government[2] sets its regional economic development goal to grow as a major hub in the Pacific Rim for tourism, information, finance, and logistics. The promotion plan adopted in 2014 identified the five new strategic industries for Busan Metro—*Marine Industry, Convergence Components & Materials, Creative Culture Industry, Bio-health Industry*, and *Knowledge Infrastructure Services Industry*. Since the second quarter of 2011, Busan's major exporting industries, Automobile Parts & Shipbuilding Materials Manufacturing, Textile and Fashion Industry, Shoe Manufacturing, and Shipbuilding, have continuously declined (Bank of Korea Busan Office 2018). Busan Metro is currently experiencing the transition from the manufacturing industries, the traditional exporting and leading industries, to the creative culture and knowledge infrastructure service industries. Martin (2011) discussed the "adaptive capability of system" and its role to self-organize a region's industrial structure during recessionary periods with growing uncertainty. The recent recession in Busan Metro's regional economy further stimulated the rapid transition towards high value-added service industries including MICE (Meetings, Incentives, Conferences, Exhibitions) industry and Film and Visual Culture industry. These are the specific target service industries within a broadly defined tourism industry. Frenken et al. (2007) argued that the development of within-sector variety during the recession resulted in the rapid recovery and growth of employment. This can be linked to the "adaptive" resilience in a regional economy that will result in positive hysteretic outcomes from recessionary shock (Martin 2011). Recently, Hartman (2016) developed "adaptive" tourism along with the coevolutionary tourism product development from a complex adaptive system (CAS). He identified strategic planning and governance as vital for enhancing adaptive capacity in a destination and, in turn, the improved adaptive capacity of a destination will further promote the variety of tourism products through related within-sector variety. Facing the growing competition in a global tourism market, a tourism destination's adaptive capacity is a key to survival, thanks to the enhanced adaptive capacity that makes the destination effectively respond to evolving local and global changes. Also, Benur and Bramwell (2015) emphasized that concentration, integration, and diversification are crucial for enhancing tourism in the competitive environment. Brouder and Eriksson (2013) and Sanz-Ibáñez and Clavé (2014) tried to apply the concept of evolutionary economic geography (EEG) to tourism research. In developing a theoretical foundation of the evolutionary reorganization and/or product diversification among tourism destinations, they borrowed the path- and place-dependent evolutionary process from EEG.

Since the early 2010s, Japan has enjoyed the remarkable growth of inbound tourism demand. Henderson (2017) analyzed the driving forces of the growth and found the six main factors for successful destination development: (1) national conditions in the country, (2) government tourism policy, (3) attractions and

---

[2]Busan Metropolitan government selected the five new strategic industries in 2014 as shown in the link below, https://english.busan.go.kr/bskeygrowth

amenities, (4) access and mobility, (5) destination marketing, and (6) international conditions. In the study, the author also found that South Korea, Taiwan, and Mainland China are the top three major markets for the growing inbound tourism in Japan. According to the inbound visitation dataset from JNTO (Japanese National Tourism Organization), over 26% of total inbound visitors to Japan were from South Korea in 2017. Kitakyushu-Fukuoka Major Metropolitan Area (Fukuoka Metro hereafter) on the island of Kyushu is the closest MMA to the Korean peninsula and attracts lots of tourists from South Korea. For 2017, the share of Korean visitors to Kitakyushu-Fukuoka MMA was 65%, 2.4 times higher than the overall share of Korean visitors to Japan (26%). Both Busan Metro and Fukuoka Metro are located far from the national capitals, but these two metros have been serving as the economic engines of the distant regions, respectively. Douglass (2013) discussed the trans-border city-region driven by Busan and Fukuoka local governments' initiative to develop a "common living sphere" by linking Busan and Fukuoka Metros beyond the national border. The author considered the decentralizing political power to local governments as a main driving force to stimulate the formation of transborder city regions for secondary cities in developed countries confronted by major economic and social transformations. Arai (2011) found that a holistic approach by experiences and changes is necessary to promote cross-border tourism between Busan and Fukuoka. Takaki and Lim (2011) compared the increasing interaction between Busan and Fukuoka to the Øresund trans-border collaboration between Denmark and Sweden. Especially, the authors analyzed the impact of the increasing interaction on tourism, logistics, and manufacturing industries. They found that the integrated cross-border business environment and industrial collaboration are the necessary steps to develop a fully integrated trans-border region in Northeast Asia. In another study by Thyne et al. (2018), researchers highlighted the importance of social distance for tourism destination development. They found a significantly different attitude towards tourism development based on the social distance from the origin countries of international visitors perceived by host residents in Japan. South Korea and Japan are closest neighbors with many cultural commonalities that can be fully utilized to link the two closest metros across the national border for developing a fully integrated trans-border region. Enhanced human connectivity through expanded tourist flows between the two metros is expected to strengthen the cohesive factors of the proposed Busan-Fukuoka trans-border region. To develop a cross-border innovation system, tourism flows will also play a vital role by facilitating knowledge transfer through the interaction and learning among actors in a trans-border region (Weidenfeld 2013).

9.3 Trend and Economic Impact Analyses

The research question of this chapter is if Busan and Fukuoka can build a cross-border megapolitan cluster by enhancing regional cohesion which will be mutually beneficial to both metros through the tourism industry. More specifically, this

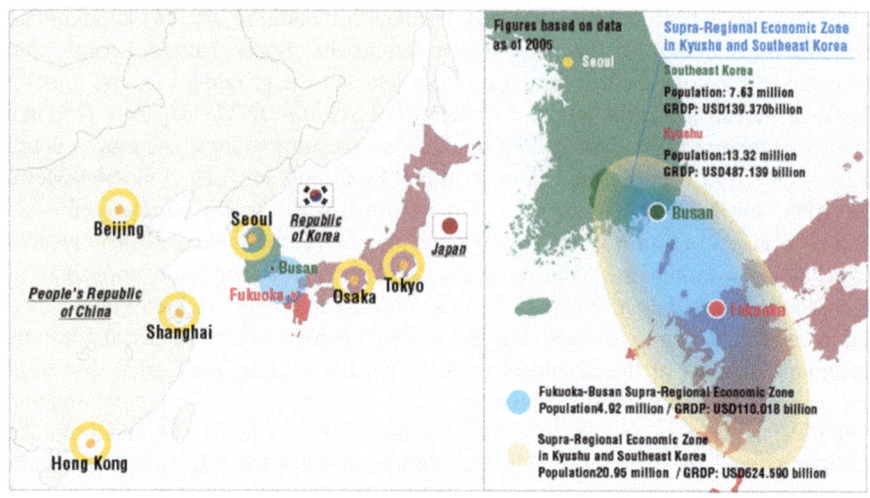

Fig. 9.2 Fukuoka-Busan Supra-regional Economic Zone (Busan-Fukuoka Megapolitan Cluster). (Source: Fukuoka – Busan Economic Cooperation Office, http://cafe.city.fukuoka.lg.jp/business/en/projects/#link01)

chapter aims to identify the recent trends of tourism industries of Busan and Fukuoka, estimate the economic impact of foreign visitor spending in Busan and Fukuoka Metros, and initiate the discussion for potential collaborative strategies to develop tourism assets for Busan-Fukuoka megapolitan cluster across the strait between Korea and Japan.

- *Study Area and Data Scale*

Busan is the second-largest metro in South Korea located in Southeast Korea, while Fukuoka is the largest city in Kitakyushu-Fukuoka MMA (Major Metropolitan Area) which is the fourth most populous MMA in Japan. As of 2017, the population of the broadly defined Busan Metro including the four neighboring cities, Ulsan, Changwon, Gimhae, and Geojeo, was 6.5 million, whereas the population of Fukuoka-Kitakyushu Greater Metropolitan Area was 5.1 million. Busan-Fukuoka megapolitan cluster across the strait between Korea and Japan is the home of 11.6 million people. When the regional boundary extends further as shown in Fig. 9.2, the population of the supra-regional economic zone in Kyushu and Southeast Korea was 20.97 million in 2017. Based on the combined population and GRDP (Gross Regional Domestic Product), Busan-Fukuoka megapolitan cluster will be the seventh-largest supra-regional economic zone in Northeast Asia, next to Tokyo, Osaka, Beijing, Shanghai, Hong Kong, and Seoul (Fig. 9.2). This will be the only supra-regional economic zone across the national border among the largest economic zones in Northeast Asia.

- *Trend Analysis & Scenario Development*

Trend analysis focuses on the growth patterns in the number of Korean visitors to Fukuoka Metro from Korea and the number of Japanese visitors to Busan

Metro for the 12-year study period (2006–2017). Additionally, we analyzed the trend in the total Korean visitors to Japan and the total Japanese visitors to Korea and the matching shares of the total inbound visitors to Japan and Korea for the same period, respectively. The trend analyses are descriptive but serve as a base for the forecast of the future trend up to the year 2023. The visitation data sources are JNTO (Japan National Tourist Organization) and the Busan Metro Government's Culture & Tourism Department.

First-order linear autoregressive (AR(1)) models are employed for the forecasting of the number of Korean visitors to Fukuoka Metro and the number of Japanese visitors to Busan Metro. With the forecast, it is possible to demonstrate the growing dependencies of the tourist flows between the two countries with the destinations at Fukuoka and Busan Metros, respectively. The estimation results lead to the discussion on the increasing importance of the potential collaboration between Busan and Fukuoka in developing cross-border tourism resources. Using the forecast results for the number of inbounding visitors to Busan and Fukuoka metros, three scenarios for economic impact analysis are developed. The assumptions in each scenario are listed below.

– *Scenario 1*
   (for the impact of Japanese visitors to Busan Metro):

   Japanese visitors to Busan Metro grow at the same growth rate from 2006 to 2017.
   Visitor spending of a Japanese visitor to Busan Metro is kept at $222.50 in 2018 USD.

– *Scenario 2*
   (for the impact of Korean visitors to Fukuoka Metro):

   Korean visitors to Fukuoka Metro grow at the same growth rate from 2006 to 2017.
   Visitor spending of a Korean visitor to Fukuoka Metro is kept at $406.53 in 2018 USD.

– *Scenario 3*
   (for the aggregated economic impact of scenarios 1 & 2):

   The number of visitors to each metro grows at the same growth rate from 2006 to 2017.
   Visitor spending of a Japanese visitor to Busan Metro is kept at $222.50 in 2018 USD, while visitor spending of a Korean visitor to Fukuoka Metro is kept at $406.53 in 2018 USD.

• *Economic Impact Estimation*
   This chapter utilizes the 2005 version of the TIIO (Transnational Interregional Input-Output) table for the estimation of the economic impact of the tourism industry in Busan and Fukuoka Metros. In 2006, Japan's IDE-JTERO (Institute of Developing Economies, Japan External Trade Organization) in partnership

with China's SIC (State Information Center) launched the TIIO project and published the 2000 version of TIIO, the predecessor of 2005 TIIO (Meng et al. 2013). The 2000 TIIO was designed to describe interregional and inter-industrial linkages in the 19 regions of East Asia, including eight subnational regions in Japan and seven subnational regions in China. Later in 2012, the TIIO table had been revised to include subnational regions of South Korea and other regions in addition to the Japanese and Chinese regions, using the 2005 dataset. Bank of Korea's Economic Statistics Department newly joined the 2005 TIIO project in 2012 (Meng et al. 2013).

The TIIO 2005 table depicts that all sectoral transactions, as well as final demand (in five categories, such as private consumption, government consumption, gross fixed capital formation, inventories, and discrepancies with the national accounts) among these regions, are fully specified. On the other hand, value-added is aggregated as the regional total and categorized into the following four types: wages and salary, operating surplus, depreciation of fixed capital, and indirect taxes less subsidies.

For the economic impact estimation, this study focused on the regions only in three East Asian countries, China, South Korea, and Japan. Also, the 20 subnational regions in the study area of the three countries are aggregated into 15 subnational regions (Table 9.1 and Fig. 9.3).

– China:

> Dongbei (C1), Huabei (C2), Huadong (C3), and Rest of China (C4 + C5 + C6 + C7).

– Japan:

> Hokkaido-Tohoku (J1 + J2), Kanto (J3), Chubu (J4), Kinki (J5), Chugoku (J6 + J7), Kyushu (J8), and Okinawa (J9).

– Korea:

> Sudokwon (K1), Jungbukwon (K2), Yeongnamkwon (K3), and Honamkwon (K4).

There are two I-O (input-output) tables in the 2005 TIIO project, one with the 10 industrial sectors and the other with 15 industrial sectors. This study used a 10-sector table (see Table 9.14 in the Appendix). I-O table was modified with the regional aggregation listed above and the Miyazawa framework, which was needed for income generation impact. Income generation impacts from foreign visitors to the study regions are estimated employing the Miyazawa framework (Miyazawa 1976), which is applied to the above 2005 TIIO. In the Miyazawa framework, consumption coefficients and value-added (wage and salary) coefficients are endogenized so that the income formation process from external changes can be dealt. As aforementioned, the 2005 TIIO table's final demand is fully specified among regions by sector, including personal consumption, our Miyazawa modification utilized only intraregional personal consumption for

Table 9.1 Regional Classification of 2005 TIIO Table

Region	Description (The regional numbers in the maps of Fig. 9.3 are in parentheses)
China (code = 5)	
C1 Dongbei	Liaoning (6), Jilin (7), Heilongjiang (8)
C2 Huabei	Beijing (1), Tianjin (2), Hebei (3), Shandong (15)
C3 Huadong	Shanghai (9), Jiangsu (10), Zhejiang (11)
C4 Huanan	Fujian (13), Guangdong (19), Hainan (21)
C5 Huazhong	Shanxi (4), Anhui (12), Jiangxi (14), Henan (16), Hubei (17), Hunan (18)
C6 Xibei	Inner Mongolia (5), Shaanxi (27), Gansu (28), Qinghai (29), Ningxia (30), Xinjiang (31)
C7 Xinan	Guangxi (20), Chongqing (22), Sichuan (23), Guizhou (24), Yunnan (25), Tibet (26)
Japan (code = 7)	
J1 Hokkaido	Hokkaido (1)
J2 Tohoku	Aomori (2), Iwate (3), Miyagi (4), Akita (5), Yamagata (6), Fukushima (7)
J3 Kanto	Ibaraki (8), Tochigi (9), Gunma (10), Saitama (11), Chiba (12), Tokyo (13), Kanagawa (14), Niigata (15), Yamanashi (19), Nagano (20), Shizuoka (22)
J4 Chubu	Toyama (16), Ishikawa (17), Gifu (21), Aichi (23), Mie (24)
J5 Kinki	Fukui (18), Shiga (25), Kyoto (26), Osaka (27), Hyogo (28), Nara (29), Wakayama (30)
J6 Chugoku	Tottori (31), Shimane (32), Okayama (33), Hiroshima (34), Yamaguchi (35)
J7 Shikoku	Tokushima (36), Kagawa (37), Ehime (38), Kochi (39)
J8 Kyushu	Fukuoka (40), Saga (41), Nagasaki (42), Kumamoto (43), Oita (44), Miyazaki (45), Kagoshima (46)
J9 Okinawa	Okinawa (47)
Korea (code = 4)	
K1 Sudokwon	Seoul (1), Incheon (2), Gyeonggi-do (3)
K2 Jungbukwon	Daejeon (4), Gangwon-do (5), Chungcheongbuk-do (6), Chungcheongnam-do (7)
K3 Yeongnamkwon	Daegu (8), Ulsan (9), Busan (10), Gyeongsangbuk-do (11), Gyeongsangnam-do (12)
K4 Honamkwon	Gwangju (13), Jeollabuk-do (14), Jeollanam-do (15), Jeju-do (16)
Others (code = 1)	
N0 Taiwan	
Q0 ASEAN5	Indonesia, Malaysia, The Philippines, Singapore, Thailand
U0 The United States	
H0 Hong Kong	
W0 Rest of the World	

consumption coefficient matrix but did not include interregional personal consumption, because foreign tourist expenditures are considered as interregional personal consumption and they should be exogenized for their impact estimations. In addition, since the value-added values are available only as a regional

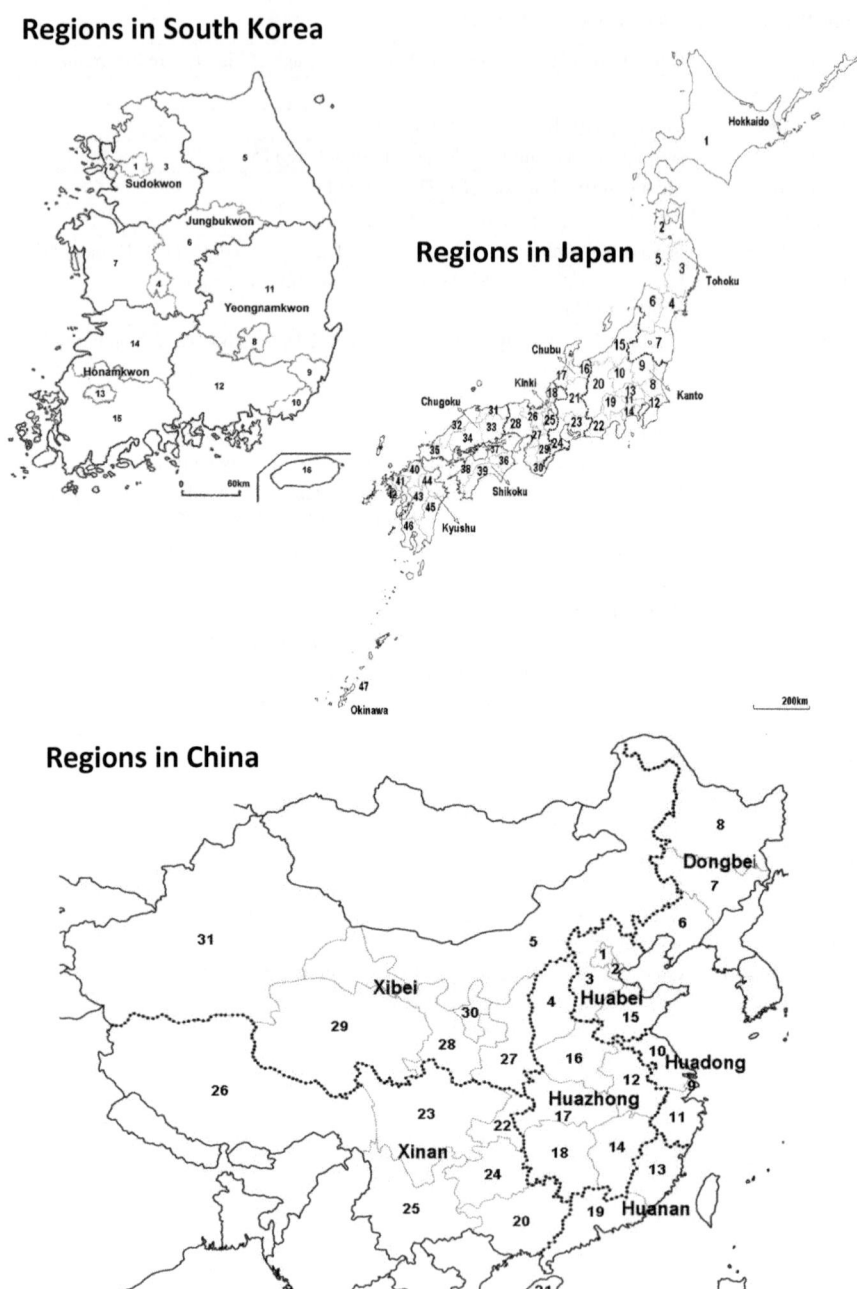

Fig. 9.3 Regional Maps of 2005 TIIO

total (a vector with one row), we modified them as a diagonal matrix and used it as a value-added coefficient matrix .

9.4   Results and Findings

This section explains the results of trend analysis and economic impact estimation of the growing tourism in the Busan-Fukuoka megapolitan cluster. Also, findings from the analytical and estimation results will be discussed.

- *Trend Analysis.*

    - *Inbound Tourism to Japan and Korea*
        For the first time, the total inbound visitors to Korea surpassed ten million in 2012 and continued to increase to reach 13.3 million in 2017. During 2017, Japan had approximately 27.4 million inbound visitors, which is over two times more inbound visitors compared to visitors to Korea. For the study period of trend analysis (2006–2017), the number of Japanese visitors to Korea has been on the rise from 2007 to 2012, mainly thanks to the strong Japanese Yen, reaching a peak of 3.52 million visitors in 2012 (see Fig. 9.4). However, due to the weak Yen and the rise of other competing destinations in Asia, the number of Japanese visitors to Korea had significantly decreased to the trough of 1.84 million in 2015. Due to the recent increase between 2015 and 2017, the total number of Japanese visitors to Korea was approximately 2.31 million, a similar level observed in 2006. On the other hand, Korean visitors to Japan had been significantly rising recently since 2011. For the 6-year period (2011–2017), the number of Korean visitors had almost quadrupled from 1.92 million in 2011 to 7.41 million in 2017. The growing popularity of Japanese tourist destinations among Korea visitors is mainly

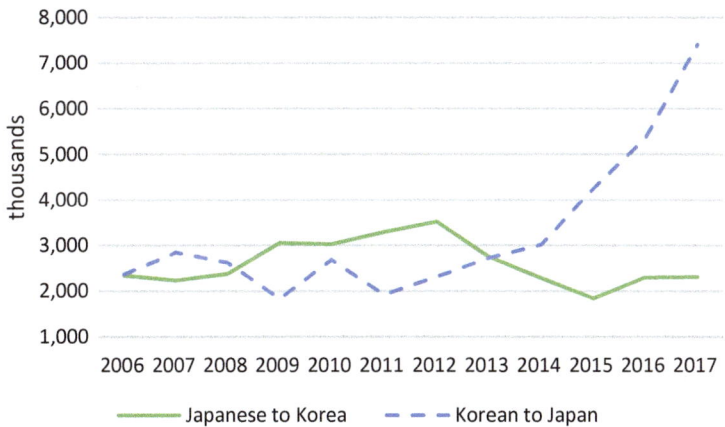

Fig. 9.4 Annual Visitor Volume between Japan & Korea (2006–2017)

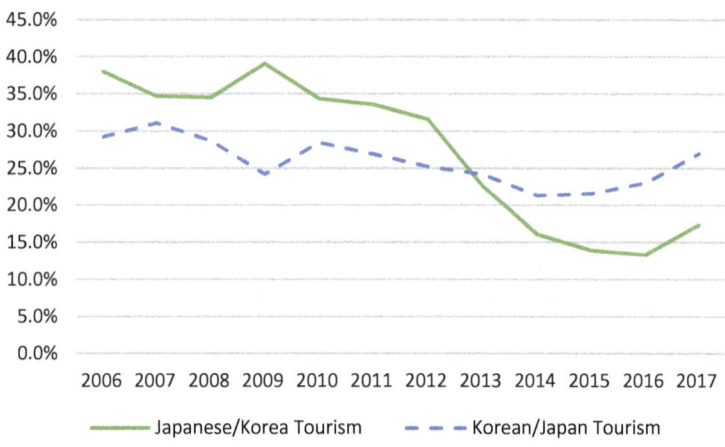

Fig. 9.5 Annual Share of Japanese (or Korea) Visitor to Korea (or Japan) (2006–2017)

attributed to the following factors: (1) lowered cost due to weak Japanese Yen to Korean Won, (2) income growth in Korea, (3) geographic proximity, (4) cultural familiarity between the two countries, and (5) successful implementation of Japanese government policy to stimulate tourism industry.

The share of Japanese visitors to Korea reached a peak of 39.1% in 2009 and continued to decline until 2016 when only 13.3% of total inbound visitors to Japan was from Korea (Fig. 9.5). A similar trend is also found for the share of Korean visitors to Japan. Since the peak of 31.1% in 2007, the share dropped by about 10% down to 21.3% in 2014. This can be explained by the rapidly increasing Chinese visitors to both countries. China became the top origin of visitors to both Japan and Korea and relative shares of Korean (or Japanese) visitors to Japan (or Korea) had decreased. But, Korea and Japan are still one of the most important origins of inbound visitors to each country. Especially, the share of Korean visitors to Japan surpassed the matching share of Japanese visitors to Korea since 2013, indicating a relatively higher importance of Korean visitors to Japan compared to Japanese visitors to Korea. Recently, both shares have rebounded.

– *Inbound Tourism to Fukuoka and Busan Metros*

As shown in Fig. 9.6, roughly 25% of total inbound visitors to Korea had visited Busan Metro from 2006 to 2013. However, the share dropped below 20% in 2014, hit the trough in 2015 (15.8%), and had stayed below 20% until 2017. In other words, only 1 out of 5 international visitors to Korea had visited Busan Metro, the second-largest metro in Korea, between 2014 and 2017. However, due to the increasing trend of total inbound visitors to Korea, the actual number of foreign visitors to Busan metro had continuously increased from 1.53 million in 2006 to 2.18 million in 2013. For 2017, the most recent year in our dataset, the total foreign visitors to Busan Metro dropped to 2.40 million. The same graph also indicates that roughly around 6% to 9%

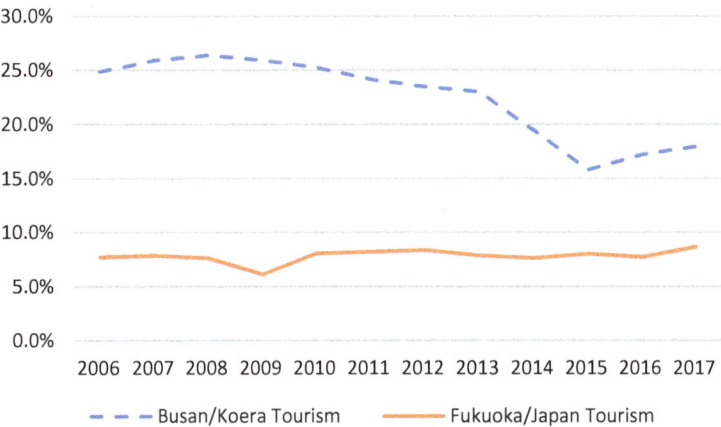

Fig. 9.6 Annual Share of Foreign Visitors to Busan (or Fukuhara) of Total Foreign Visitors (2006–2017)

of annual foreign visitors to Japan had visited Fukuoka Metro between 2006 and 2017. The relative share of Fukuoka Metro as a tourism destination in Japan among foreign visitors is quite stable over the study period. The huge gap between the two metros might be due to the relative importance of each metro and the competition among other tourism destinations in each country. Especially, Busan Metro and its neighboring cities have been the center of heavy and manufacturing industries in Korea and industrial activities might have attracted many business-related visitors to Busan Metro. Fukuoka Metro, even with the distant location from Tokyo and other major metro areas like Osaka, could maintain its share thanks to the automobile-related industry and other surging industries with the start-ups in biotechnology and semiconductor-related industries.

Of the total foreign visitors to Fukuoka Metro, Korea, visitors' share ranged from 50% to 65% and it peaked at 70.6% in 2007 (Fig. 9.7). The absolute majority of foreign visitors to Fukuoka Metro is clearly from Korea. On the other hand, the share of Japanese visitors to Busan out of total foreign visitors to Busan ranged from 17% and 25%. It is important to notice here that Fukuoka Metro fully utilizes its geographic proximity to Korea, while Busan Metro fails to attract tourists from its closest foreign country.

– *Forecast for Japanese/Korean visitors to Busan/Fukuoka*

From the peak, 635,000, Japanese visitors to Busan, the number of Japanese visitors has declined to 471,000 visitors in 2017. Based on the recent decline, Japanese visitors are expected to decline by 0.3% annually on average, leading to about 460,000 visitors by 2025. Conversely, Korean visitors to Fukuoka had increased by over 270% from 413,000 visitors in 2006 to 1,534,000 visitors in 2017. This trend is expected to slow down due to the recent tension between Japan and Korea governments on trade, national

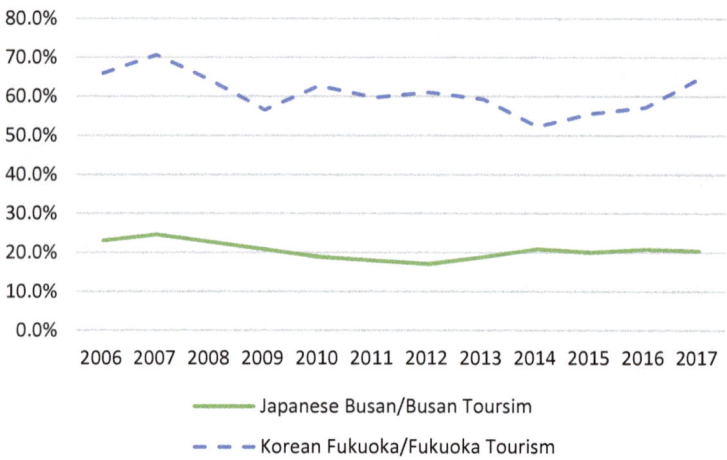

Fig. 9.7 Annual Share of Foreign Visitors to Busan (or Fukuhara) of Total Foreign Visitors (2006–2017)

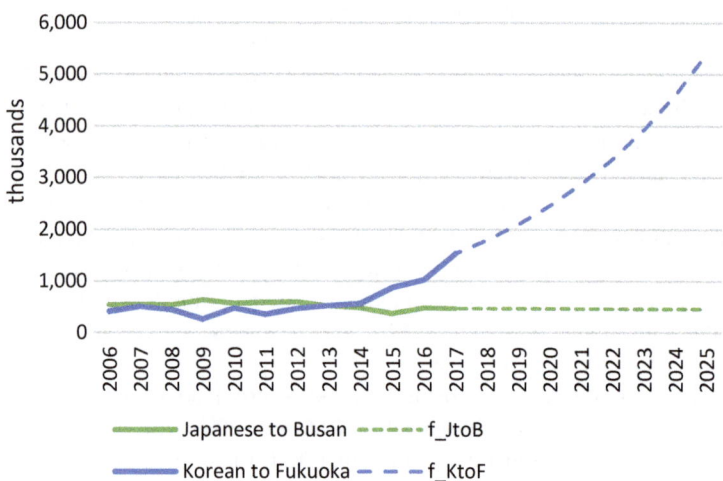

Fig. 9.8 Annual Visitor Trend & Forecast to Busan (or Fukuoka) from Japan (or Korea) (2006–2025)

security, and history-related issues. However, if this trend persists, the total Korean visitors to Fukuoka are expected to reach over 5.4 million by 2025 (see Fig. 9.8).

The relative importance of Fukuoka Metro as a rising tourism destination among Korean visitors can be also confirmed from the rapid growth of air passengers and increasing flight services between Gimhae International Airport (PUS) in Busan Metro and Fukuoka International Airport (FUK) in Fukuoka Metro (Fig. 9.9). Back in 2003, there were 846 air passenger flights

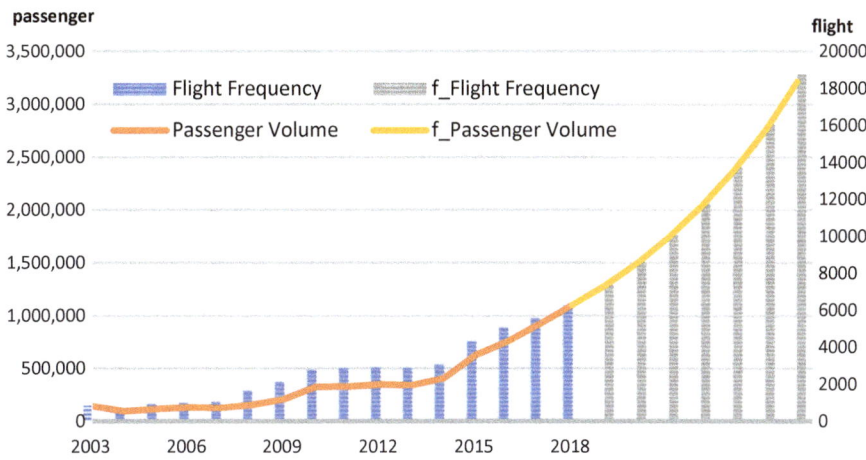

Fig. 9.9   Annual Air Passenger and Flight Trend & Forecast between PUS and FUK (2006–2025)

between the two airports annually, but the number of air passenger flights had increased by 62.8% to 1377 flights by 2010. The number of air passenger flights had skyrocketed to 6296 by 2018, 4.6 times higher compared to the 2010 level. Again, with the recent tension between the two countries and the saturation in the growing demand of Korean visitors to Fukuoka, the growth is expected to slow. But, if the recent trend continues, the average annual growth will be about 16.8%, leading to 18,770 air passenger flight services by 2025. The number of air passengers shows a similar growth pattern from 140,000 passengers in 2003 to 1,100,000 passengers in 2008. If this growth continues, there will be 3,210,000 air passengers between PUS and FUK by 2025. The growing demand for cruise services and ferries linking Port of Busan in Busan Metro and Hakata Port in Fukuoka Metro.

The following section summarizes the annual economic impacts of tourists' spending in Busan Metro and Fukuoka Metro, estimating with the 2005 TIIO table among three countries, Japan, Korea, and China for the study period between 2013 to 2018 and the forecast period of 2018–2023.

- *Economic Impact Estimation*

This study employed the 2005 TIIO table for the annual economic impacts of Korean (or Japanese) tourists' spending in Fukuoka (or Busan) Metro between 2013 and 2018. While the estimated annual impact from 2013 to 2018 is based on the observed visitation dataset and the average spending per capita in 2018, the annual economic impacts from 2019 to 2023 are based on the visitation forecasts from the previous section.

Economic impacts are measured in terms of output and income for three countries in the model at the national level, followed by the regional economic impact for two regions with direct impact from tourists' spending, namely Yeongnamkwon of Korea and Kyushu of Japan (see Table 9.1 and Fig. 9.3 for

Table 9.2 Total National Economic Impact of Japanese Visitors to Busan Metro (in 2018 USD)

	2013	2018	2023
Total output impact (China)	10,914,705	9,900,261	9,750,929
Total output impact (Japan)	13,351,131	12,110,238	11,927,571
Total output impact (Korea)	508,931,205	461,629,675	454,666,609
Total income impact (China)	1,458,935	1,323,338	1,303,377
Total income impact (Japan)	3,252,833	2,950,505	2,906,001
Total income impact (Korea)	133,105,569	120,734,355	118,913,239

Source: Estimation by authors with 2005 TIIO table (annual impact for all years between 2013 and 2023 is available upon request)

Table 9.3 Total Regional Economic Impact of Japanese Visitors to Busan Metro (in 2018 USD)

	2013	2018	2023
Total output impact (Kyushu region)	1,169,106	1,060,446	1,044,450
Total output impact (Yeongnamkwon region)	398,698,062	361,641,918	356,187,033
Total income impact (Kyushu region)	302,371	274,268	270,131
Total income impact (Yeongnamkwon region)	110,576,220	100,298,948	98,786,073

Source: Estimation by authors with 2005 TIIO table (annual impact for all years between 2013 and 2023 is available upon request)

details). Yeongnamkwon region (K3) is the home of the Busan Metro in this study, while the Kyushu region (J8) is the home of Fukuoka Metro. For each study region (or nation), aggregated output and income impacts are summarized for each scenario.

– *Scenario 1: Economic impact of Japanese visitors to Busan Metro*

Scenario 1 estimated the annual economic impact of Japanese visitors to Busan Metro, and Table 9.2 shown below summarizes the nationwide total output and income impacts for China, Japan, and Korea for selected years during the study period (2013–2023). As expected, the largest impact is in Korea, followed by the second largest output impact in Japan and the least economic impact in China. Interestingly, with the growing inter-industrial linkages between Korea and China, the gap in total output impact between Japan and China is not huge, only 22.3% higher in Japan than in China. However, due to the higher wage level in Japan, the gap in income impact between Japan and China is much larger, the total income impact in Japan is 2.23 times higher than that in China. With the decreasing trend of Japanese visitors to Busan by 2023, the economic impact is expected to decrease, as well.

Table 9.3 below summarizes total output and income impacts for Kyushu (J8) and Yeongnamkwon (K3) regions for selected years during the study

Table 9.4 Total National Economic Impact of Korean Visitors to Fukuoka Metro (in 2018 USD)

	2013	2018	2023
Total output impact (China)	3,110,547	10,623,278	23,331,850
Total output impact (Japan)	309,747,159	1,057,862,053	2,323,376,952
Total output impact (Korea)	1,188,654	4,059,544	8,915,956
Total income impact (China)	412,838	1,409,941	3,096,647
Total income impact (Japan)	93,326,321	318,732,137	700,029,743
Total income impact (Korea)	188,484	643,720	1,413,798

Source: Estimation by authors with 2005 TIIO table (annual impact for all years between 2013 and 2023 is available upon request)

period (2013–2023). The total output impact in Yeongnamkwon (K3) of Korea, a destination of Japanese tourists in scenario 1 was $361.6 million in 2018 and expects to decline slightly to $356.2 million by 2023 due to the decrease in the number of Japanese visitors to Busan. Kyushu region of Japan, a neighboring metro in the Busan-Fukuoka megapolitan cluster also receives output impact but very negligible at $1.1 million in 2018 and $1.0 million in 2023. A similar pattern holds for income impact between the two metros, i.e. inter-regional inter-industrial ties through the tourism industry and/or more broadly defined service industry between the two metros are very limited. In 2018, of the nationwide total output impact in Korea, 78.3% ($361.6 million) is in the Yeongnamkwon (K3) region, which is composed of $191.0 million of direct impact in Busan Metro and $170.6 million of indirect and induced impact in Yeongnamkwon including Busan Metro.

– *Scenario 2: Economic impact of Korean visitors to Fukuoka Metro*

Scenario 2 measures the annual economic impact of Korean visitors to Fukuoka Metro as summarized in Table 9.4 for the nationwide total output and income impacts for China, Japan, and Korea in the three selected years during the study period (2013–2023). It is natural to have the largest impact in Japan from the Korean visitors' spending in Fukuoka (direct impact in Fukuoka Metro), the second-largest output impact in China, and the least economic impact in Korea. This pattern is the opposite of the results in scenario 1, in which output impact was the largest in the destination country (Korea), followed by the origin country (Japan), and lastly China. But, in scenario 2, China has a higher output impact than Korea, an origin country and this might be due to the higher inter-industrial linkages between Japan and China than the ties between Japan and Korea. According to WTO (World Trade Organization), Japan was the third-largest export market of China (6.2% of total Chinese export), whereas Korea was the fourth largest export market (4.5% of total Chinese export). With the rapidly growing visitation of Koreans to Fukuoka Metro, the economic impact is expected to grow by 220% for the 5-year period (2018–2023).

Table 9.5 lists total output and income impacts for Kyushu (J8) and Yeongnamkwon (K3) regions for years 2013, 2018, and 2023 under scenario

Table 9.5 Total Regional Economic Impact of Korean Visitors to Fukuoka Metro (in 2018 USD)

	2013	2018	2023
Total output impact (Kyushu region)	237,761,520	812,013,546	1,783,421,148
Total output impact (Yeongnamkwon region)	450,508	1,538,596	3,379,210
Total income impact (Kyushu region)	75,224,385	256,909,611	564,249,248
Total income impact (Yeongnamkwon region)	58,374	199,362	437,857

Source: Estimation by authors with 2005 TIIO table (annual impact for all years between 2013 and 2023 is available upon request)

2. The total output impact in Kyushu (J8) of Japan, a destination of Korean tourists in scenario 2, was $237.8 million in 2013, roughly 40% lower than the output impact of $398.7 million from Japanese visitors to Busan (in scenario 1). However, with the rapid growth of Korean visitors and the higher per-visitor spending of Korean visitors in Fukuoka Metro during 2018, the economic impact of Korean visitors in Kyushu (J8) became 2.25 times larger than that of Japanese visitors in Yeongnamkwon (K3). The output impact got shifted for the first time in 2015 (based on the annual impact estimation results[3]) and the gap in output impact had continuously increased. In 2018, of the nationwide total output impact in Japan, 76.8% ($812.0 million) is in the Kyushu (J8) region, which is composed of $399.6 million of direct impact in Busan Metro and $412.4 million of indirect and induced impact in Kyushu including Fukuoka Metro. An interesting difference between scenario 1 and 2 is the larger multiplier effect under scenario 2 than in scenario 1. In other words, the ripple effects of Korean visitors to Fukuoka are larger than that of Japanese visitors to Busan Metro both at national and regional levels. This confirms that the tourism industry in Japan has a higher multiplier impact than that in Korea.

– *Scenario 3: Aggregated economic impact of Scenarios 1 & 2*

Scenario 3 is the sum of scenarios 1 and 2. Under scenario 3, the estimated economic impact includes the aggregated impact of Japanese visitors to Fukuoka Metro (scenario 1) and that of Korean visitors to Busan Metro (scenario 2). As summarized in Table 9.6, the output and income impact of the two opposite flows of visitation to Busan and Fukuoka metros in Korea was larger in 2013, but the combined effect became larger for Japan in 2018 and the gap had widened and expected to increase until 2023 (see Figs. 9.10 and 9.11). In scenario 3, the share of direct economic impact in Japan ranges from 36% to 38% of the total economic impact, whereas the share of direct economic impact in Korea ranges from 40% to 42% of the total economic

---

[3]The annual impact estimation is not included in this chapter; however, the annual results will be available from the authors upon request.

Table 9.6 Total National Economic Impact of Aggregated Korean & Japanese Visitors to Fukuoka & Busan Metros (in 2018 USD)

	2013	2018	2023
Total output impact (China)	14,025,253	20,523,539	33,082,779
Total output impact (Japan)	323,098,290	1,069,972,291	2,335,304,523
Total output impact (Korea)	510,119,859	465,689,219	463,582,565
Total income impact (China)	1,871,773	2,733,279	4,400,024
Total income impact (Japan)	96,579,155	321,682,642	702,935,744
Total income impact (Korea)	133,294,053	121,378,074	120,327,037

Source: Estimation by authors with 2005 TIIO table (annual impact for all years between 2013 and 2023 is available upon request)

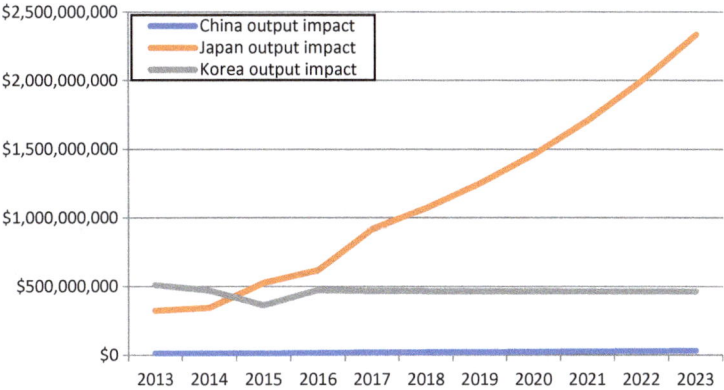

Fig. 9.10 National Outcome Impact of Aggregated Korean & Japanese Visitors to Fukuoka & Busan Metro (in 2018 USD)

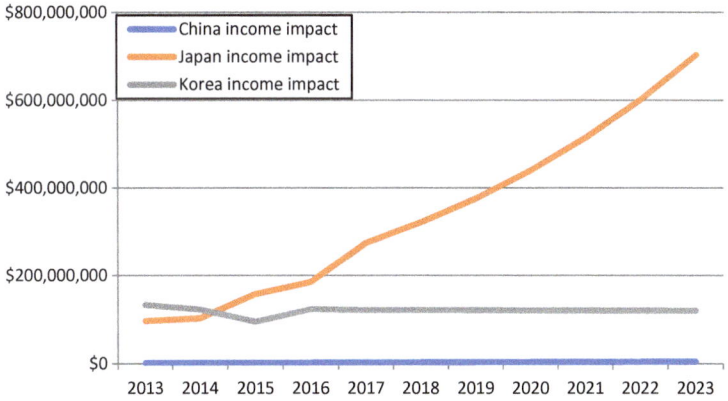

Fig. 9.11 National Income Impact of Aggregated Korean & Japanese Visitors to Fukuoka & Busan Metro (in 2018 USD)

Table 9.7 Total Regional Economic Impact of Aggregated Korean & Japanese Visitors to Fukuoka & Busan Metros (in 2018 USD)

	2013	2018	2023
Total output impact (Kyushu region)	238,930,625	813,073,991	1,784,465,598
Total output impact (Yeongnamkwon region)	399,148,571	363,180,514	359,566,244
Total income impact (Kyushu region)	75,526,756	257,183,879	564,519,379
Total income impact (Yeongnamkwon region)	110,634,595	100,498,310	99,223,931

Source: Estimation by authors with 2005 TIIO table (annual impact for all years between 2013 and 2023 is available upon request)

impact. This indicates that the multiplier impact of the tourism industry in Japan is higher than that in Korea through denser industrial linkages in Japan. The aggregated economic impact in Japan had started to exceed the economic impact in Korea since 2015 and the gap has widened until 2018. If the trend observed between 2006 and 2018 persists, the aggregated output impact in Japan will be 5 times bigger than the aggregated output in Korea by 2023, because of the increased number of Korean travelers to Fukuoka. The aggregated income impact in Japan will be 5.8 times higher than that in Korea by 2023. Both outcome and income impacts in China are negligible compared to the impact in Korea or Japan.

In Table 9.7, the regional economic impact of the aggregated Korean and Japanese visitors to Fukuoka and Busan Metros is presented for Kyushu (J8) and Yeongnamkwon (K3) regions, respectively. The same patterns found in the national economic impact holds at the regional level. Under scenario 3, the total output impact in Kyushu (J8) was $813.1 million in 2018, about 55% lower than the output impact of $363.2 million in Yeongnamkwon (K3). The gap further increases from 2018 to 2023 as shown in Figs. 9.12 and 9.13.

- *Impact on Regional Industries*

  For the regional economic impact comparisons, this study ranked the top regional industries with the highest indirect impact on output from the visitor spending and the total regional income increase from the visitor spending in destinations.

  – *Regional Industries with the Highest Indirect Impact on Output*

    Of the 10 sectors (listed in Table 9.14 of Appendix) in 15 subnational regions, the "*Services*" industry in a destination of tourists has a direct impact from visitor spending. On the other hand, the indirect impact can be estimated for regional industries that benefit indirectly through inter-regional inter-industrial linkages. Tables 9.8, 9.9, and 9.10 shown below list the top five

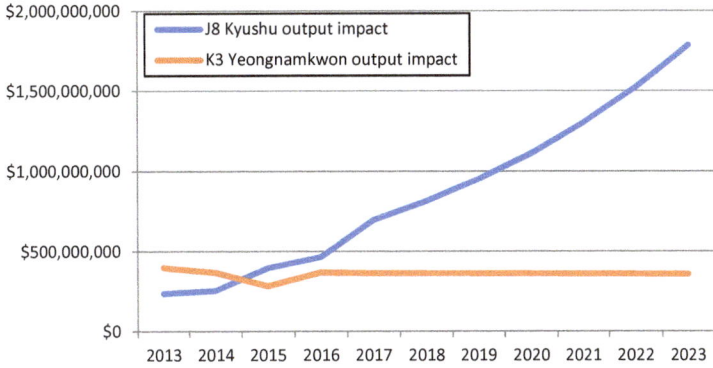

Fig. 9.12 Regional Outcome Impact of Aggregated Korean & Japanese Visitors to Fukuoka & Busan Metro (in 2018 USD)

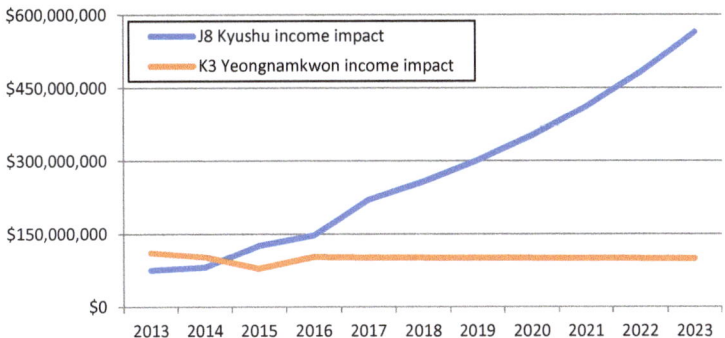

Fig. 9.13 Regional Income Impact of Aggregated Korean & Japanese Visitors to Fukuoka & Busan Metro (in 2018 USD)

Table 9.8 Top 5 Regional Sectors with High GRP Increase from Indirect Impact in Scenario 1 (2018)

Ranking	Region	Sector	Increase (2018 USD)
1	Yeongnamkwon (Korea)	Services	88,047,678
2	Sudokwon (Korea)	Services	35,618,200
3	Yeongnamkwon (Korea)	Basic industrial materials	20,390,442
4	Yeongnamkwon (Korea)	Household consumption products	15,497,132
5	Yeongnamkwon (Korea)	Processing and assembling	11,799,713

Note: "Basic Industrial Materials" include Printing and Publishing, Drugs and Medicine, and Plastic Products. "Household Consumption Products" include Beverage, Food Products, and Wearing Apparel. "Processing and Assembling" include Electronic Products and Household Electronical Products

Table 9.9 Top 5 Regional Sectors with High GRP Increase from Indirect Impact in Scenario 2 (2018)

Ranking	Region	Sector	Increase (2018 USD)
1	Kyushu (Japan)	Services	245,429,271
2	Kanto (Japan)	Services	56,234,163
3	Kyushu (Japan)	Trade	40,557,466
4	Kyushu (Japan)	Household consumption products	31,303,313
5	Kanto (Japan)	Basic industrial materials	25,736,548

Note: "Basic Industrial Materials" include Printing and Publishing, Drugs and Medicine, and Plastic Products. "Household Consumption Products" include Beverage, Food Products, and Wearing Apparel. "Trade" includes Wholesale and Retail Trades

Table 9.10 Top 5 Regional Sectors with High GRP Increase from Indirect Impact in Scenario 3 (2018)

Ranking	Region	Sector	Increase (2018 USD)
1	Kyushu (Japan)	Services	245,755,919
2	Yeongnamkwon (Korea)	Services	88,223,342
3	Kanto (Japan)	Services	57,797,269
4	Kyushu (Japan)	Trade	40,668,332
5	Sudokwon (Korea)	Services	36,166,096

Note: Trade includes Wholesale and Retail Trades

regional industries with the largest indirect impact in 2018 measured by GRP (Gross Regional Product) increases for scenarios 1, 2, and 3, respectively.

In scenario 1, the regional industry with the highest GRP increase is the "*Services*" industry in Yeongnamkwon (K3) of Korea with $88.1 million as the indirect impact. The second-largest indirect impact was captured in the "Services" industry of Sudokwon (K1) of Korea, equivalent to Seoul Metropolitan Area with the largest regional economy in South Korea. The "*Services*" industry of Sudokwon gains about 40.5% of the indirect impact in Yeongnamkwon's "*Services*" industry. Other top regional industries that benefit from the Japanese visitor spending in Busan Metro are all located in Yeongnamkwon—"*Basic Industrial Materials*," "*Household Consumption Products*," and *Processing and Assembling*." The total output multiplier in 15 subnational regions in scenario 1 is 2.53, indicating every $100 a Japanese tourist spends in Busan Metro generates an additional $153 in the 15 subregional regions of this study. When the study area is confined to Yeongnamkwon, the total output multiplier reduces to 1.89, indicating that not all the indirect impact is kept locally, instead, some are lost to the other regions through industrial linkages. The share of indirect impact kept locally in Yeongnamkwon is 58.3%.

In the case of scenario 2, the share of indirect impact kept locally in Kyushu is slightly higher at 61.3% compared to scenario 1. This represents that the higher share of the ripple effect from Korean visitor spending in Fukuoka

Metro is kept in Kyushu. Similar to the case of scenario 1, the second-highest GRP increase from indirect impact under scenario 2 is with the "*Services*" industry of the Kanto region (J3) containing the national capital of Tokyo and the 10 other surrounding prefectures near Tokyo (Table 9.9). The indirect impact captured by Kanto's "*Services*" industry was only 22.9% of the indirect impact in the "*Services*" industry of Kyushu. Other higher-ranked regional industries include two in Kyushu ("*Trade*" and "*Household Consumption Products*") and one in Kanto ("*Basic Industrial Materials*"). The overall output multiplier in 15 subnational regions across three countries is 2.68, while the local output multiplier for Kyushu is 2.03 in scenario 2. Both multipliers are larger than the multipliers found in scenario 1. Every $100 Korean visitor's spending generates an additional $168 across the overall study area, of which $103 additional output is kept locally in Kyushu. In other words, the leakage of indirect impact on other regions' industries in scenario 2 is lower than that in scenario 1. This can be partly explained by denser local inter-industrial linkages with the "Services" industry in Kyushu than those linkages in Yeongnamkwon.

Table 9.10 lists the top five regional industries from the aggregated indirect output impact in scenario 3. Due to a much higher direct impact from Korean visitor spending in Fukuoka Metro than from Japanese visitor spending in Busan Metro, the "*Services*" industry in Kyushu ranked at the top with the highest GRP increase from indirect impact in scenario 3, followed by Yeongnamkwon's "*Services*" industry. Almost two-thirds of the indirect output impact of Yeongnamkwon's "*Services*" industry is found in Kanto's "*Services*" industry. Kanto is over 1000 km away from Fukuoka Metro and not a destination of visitors in the scenarios, whereas both Kyushu and Yeongnamkwon contain the tourists' destinations in this study. Kanto's "*Services*" industry benefit from the Korean visitor spending in Fukuoka Metro is mainly driven by the solid inter-regional industrial linkages with Kyushu's "*Services*" industry. Ranked at fourth, Kyushu's "Trade" has $40.7 million of indirect output impact, followed by Sudokwon's "Services" industry with $36.2 million.

Scenario 2, the share of indirect impact kept locally in Kyushu is slightly higher at 61.3% compared to scenario 1. This represents that the higher share of the ripple effect from Korean visitor spending in Fukuoka Metro is kept in Kyushu. Similar to the case of scenario 1, the second-highest GRP increase from indirect impact under scenario 2 is with the "*Services*" industry of the Kanto region (J3) containing the national capital of Tokyo and the 10 other surrounding prefectures near Tokyo. The indirect impact captured by Kanto's "*Services*" industry was only 22.9% of the indirect impact in the "*Services*" industry of Kyushu. Other higher-ranked regional industries include two in Kyushu ("*Trade*" and "*Household Consumption Products*") and one in Kanto ("*Basic Industrial Materials*"). The overall output multiplier in 15 subnational regions across three countries is 2.68, while the local output multiplier for Kyushu is 2.03 in scenario 2. Both multipliers are larger than the multipliers

Table 9.11 Top 5 Regions with the High Income Increase from Total Impact in Scenario 1 (2018)

Ranking	Region	Income increase (2018 USD)
1	Yeongnamkwon (Korea)	100,298,948
2	Sudokwon (Korea)	16,337,215
3	Jungbukwon (Korea)	2,295,287
4	Honamkwon (Korea)	1,802,904
5	Kanto (Japan)	1,249,033

Table 9.12 Top 5 Regions with the High Income Increase from Total Impact in Scenario 2 (2018)

Ranking	Region	Income increase (2018 USD)
1	Kyushu (Japan)	256,909,611
2	Kanto (Japan)	31,895,124
3	Kinki (Japan)	11,736,746
4	Western Japan (Japan)	7,736,358
5	Chubu (Japan)	6,728,575

Table 9.13 Top 5 Regions with High Income Increase from Total Impact in Scenario 3 (2018)

Ranking	Region	Income increase (2018 USD)
1	Kyushu (Japan)	257,183,879
2	Yeongnamkwon (Korea)	100,498,310
3	Kanto (Japan)	33,144,157
4	Sudokwon (Korea)	16,640,965
5	Kinki (Japan)	12,362,687

found in scenario 1. Every $100 Korean visitor's spending generates an additional $168 across the overall study area, of which $103 additional output is kept locally in Kyushu. In other words, the leakage of indirect output impact on other regions' industries in scenario 2 is lower than that in scenario 1. This can be partly explained by denser local inter-industrial linkages with the "Services" industry in Kyushu than those linkages in Yeongnamkwon.

- *Regions with the Highest Total Income Impact*

Increases of total regional income are compared among the 15 subnational regions and listed in Tables 9.11, 9.12, and 9.13 for scenarios 1, 2, and 3, respectively. The income increases in these tables include direct and indirect income impacts from visitor spending for the year 2018.

The largest income impact is from the tourists' destination which has a direct impact from visitor spending. In scenario 1, Yeongnamkwon, the home of Busan Metro, has the highest income impact among the 15 subnational regions (Table 9.11). Every $100 spending by Japanese visitors in Busan Metro generates $53 of total income in Yeongnamkwon in 2018. Sudokwon is ranked second, but the total income impact is only about 16.3% of the income impact in Yeongnamkwon. Other regions with a high income impact include Jungbukwon (K2) and Hanamwon (K4) of Korea and interestingly

Kanto (J3) of Japan. In other words, every $1000 Japanese visitor spends induced approximately $7 of income increase in the Kanto region of Japan. About 80.2% of the total income impact is kept locally in Yeongnamkwon, followed by 13.1% in Sudokwon of Korea. While almost all (96.6%) of total income impact is in the four regions of Korea, Japan also has 2.4% of the total income impact. Among the seven subnational regions in Japan, Kanto (J3) and Kinki (J5) regions generate 1.0% and 0.5% of total income impact, respectively. The closest Japanese region of Busan Metro, the destination of tourists in scenario 1, receives only 0.2% of total income impact. It is evident that the industrial linkages of the tourism industry in Busan Metro with the related industries in Kyushu is not strong enough to result in a tangible ripple impact in Kyushu.

Under scenario 2, Kyushu has the highest total income impact, as expected (Table 9.12). Every $100 spending by Korea visitor in Fukuoka Metro generates $64 of total income in Kyushu, about 22.4% higher total impact in a tourism destination compared to scenario 1. Again, the capital area, Kanto is ranked second, but the total income impact is only about 12.4% of the income impact in Kyushu. Other regions with the high income impact are all in Japan, Kanto (J3), Kinki (J5), Western Japan (J6 + J7), and Chubu (J4), whereas the combined GRPs of Kanto, Kinki, and Chubu is about three-quarters of the Japanese national GDP. None of the Korean regions is ranked high. Approximately 80.1% of the total income impact is kept locally in Kyushu. The absolute majority of total income impact (99.4%) is kept domestically in Japan, whereas only 0.2% of total income impact is in Korea, an even smaller share than 0.4% in China. Among the subnational Korean regions, Sudokwon (K1) gains 0.1%, followed by Yeongnamkwon (K3) region with 0.06%. Yeongnamkwon is the closest region of Fukuoka Metro, the destination of tourists in scenario 2 but it has not benefitted from the increasing tourists spending in Fukuoka Metro. Again, this confirms the limited industrial linkages of Yeongnamkwon's industries to Kyushu's tourism industry.

Table 9.13 lists the top regions with the highest income impact from the combined visitor spending in scenarios 1 and 2. Two destinations in the scenarios are ranked first and second with $257.2 million of income impact in Kyushu and $100.5 million of income impact in Yeongnamkwon, respectively. Even $100 spent by either a Korean (or Japanese) visitor in Fukuoka (of Busan) Metro induced $44 of income impact in Kyushu, followed by $17 in Yeongnamkwon. Other regions also receive an increased income impact through inter-regional inter-industrial linkages, indirectly. These regions include the Kanto region of Japan, the Sudokown region of Korea, and the Kinki region of Japan. These results illuminate the fact that both countries have the centralized economic system (interindustry relationships), in which two respective capital regions (Sudokwon in Korea and Kanto in Japan) receive relatively large indirect and induced impacts from tourism activities in peripheral regions (Yeongnamkwon in Korea and Fukuoka (Kyushu) in Japan). Hence, successful tourism development of peripheral regions will benefit not only concerned regions but also other regions, especially core (capital) regions, in these two countries.

9.5 Conclusion

Using the historic annual visitor data, this study analyzed the recent trends of inbound visitors to Busan and Fukuoka Metros and estimated the economic impact of Japanese (or Korean) visitor spending in Busan (or Fukuoka) Metro. Trend analysis clearly indicates the rapid growth in the number of Korean visitors to Fukuoka since 2011, whereas the growth in the number of Japanese visitors is sluggish and even on a decline since the early 2010s. With the recent surge of Chinese visitors to Japan and Korea, the relative shares have declined, but these two neighboring countries still exchange a large number of visitors. Before 2013, the number of Japanese visitors to Busan Metro had been larger than the number of Korean visitors to Fukuoka Metro. However, this trend had reversed since 2013 and the forecast indicated that at least eight times more Korean visitors are expected to visit Fukuoka than Japanese visitors to Busan. The rapid growth of the tourism industry in Fukuoka Metro has been stimulated by the Japanese Central government policy to revitalize regional economies with proper investment for tourism destination development (Andonian et al. 2016). Conversely, Busan Metro, the second largest only next to Seoul Metropolitan Area, and closest neighbor to Japan has continuously failed to attract Japanese visitors recently. The tourism industry is one of the key target industries for Busan Metro's economic development strategies for the post-manufacturing era. However, Busan and Southeast Korea do not fully utilize the relative advantage in geographic proximity and better accessibility in attracting visitors from Japan, especially from Kyushu through Hakata Port and Fukuoka International Airport. Two metros in the *Øresund* region, a cross-border region between Sweden and Denmark, have successfully created a functioning supra-regional economic zone and the tourism industry plays a vital role in the regions' economic growth. Megapolitan cluster in the U.S. (Nelson and Lang 2011) can be another benchmark for Busan-Fukuoka Cluster. The key to the creation of a functioning megapolitan cluster is the strengthened cohesion within the cluster by linking multiple metros with the expanded flows of people, goods and services, and knowledge. It is always desirable to have mutually beneficial exchanges between the metros for the sustainable growth of a functioning megapolitan cluster. The tourism industry is a good channel for the exchange of people, goods and services, and knowledge. The unbalanced growth of tourism industries of the Busan-Fukuoka megapolitan cluster in recent years cannot guarantee the sustainable growth of the cluster in the long-run. Even the rapidly growing tourism industry in Fukuoka Metro is vulnerable to external shocks if it heavily depends on a single origin country. The economic impact analyses show that the limited inter-industrial linkages of the tourism industry between Kyushu and Yeongnamkwon, the two closest cross-border neighbors between Korea and Japan.

Busan and Fukuoka Metro governments have been working together for the successful launch of the Fukuoka-Busan Supra-Regional Economic Zone; however, they have not shown any tangible results yet. One of the low hanging fruits can be tourism destination development. Especially, Busan Metro can learn lessons from the successful experience in Fukuoka Metro and other nonmetropolitan areas in Kyushu. Moreover, two cross-border metros should initiate a set of strategies for tourism destination development collaboratively. For instance, oceanic sports can be a good example to attract more inbounding visitors from diversified origin countries including U.S. and E.U. states. By cohosting international oceanic sports events in Busan-Fukuoka megapolitan cluster, tourists from various origin countries can have opportunities to visit both metros in two countries in Northeast Asia. Additionally, by codeveloping and marketing various tourism resources such as islands between two metros, Busan-Fukuoka megapolitan cluster can be recognized as the Mediterranean in Northeast Asia and grows as a destination for increasing demand for cruise tourism in Asia, especially among Chinese. Lastly, both Busan and Fukuoka Metros are on the bullet train networks in each country. Multimode travel packages for domestic travelers can be extended to include Busan for Japanese travelers and Fukuoka for Korean travelers. Two metro governments are required to build a comprehensive understanding of the travelers' behavior in each destination, including major activities, visitor spending by categories, level of satisfaction, other basic demographic statistics, etc. Busan-Fukuoka tourism bureau can be formed and it can develop a common platform and survey instruments to collect information from visitors. A good example is LVCVA (Las Vegas Convention & Visitor Authority) which performs annual visitor profile study for tourism product development, marketing, and research purposes.

Facing growing uncertainty in the complex world, regional collaboration for tourism resource development can lower the vulnerability of the Busan-Fukuoka Megapolitan tourism industry to unexpected external shocks. Recently, the escalating political tension between Japan and Korea stemming from the different views on the sad history between the two countries in the early twentieth century and the worldwide pandemic of COVID-19 are the good examples of unexpected events. Unfortunately, with the escalating tension at the national level, many of the smaller regional tourism destinations in Kyushu and Yeongnamkwon have suffered a lot due to the significant loss of visitors. However, when these two metros share collaboratively developed tourism resources at a regional level, the cross-border region between Korea and Japan can work together to minimize the shock from the international tension. Also, with COVID-19, people tend to travel a shorter distance to minimize potential exposure to health risks. Travels within the Busan-Fukuoka cluster do not require long-distance trips, but travelers can still enjoy the exotic tourism destinations encompassing two nations with different cultures. Consequently, the collaborative development of tourism resources in the Busan-Fukuoka megapolitan cluster will benefit both metros by sustainably managing negative shocks from the unexpected events and enhancing the resilience of the tourism industry in the cluster.

Like many other studies, this study has limitations and the issues that can be addressed in future research. In the lack of itemized visitor spending data, this study

could not detect a more detailed economic impact, e.g., spending for shopping could
have been separated to measure the impact from retail trade rather than from the
aggregated service industries. Also, due to the lack of regional-level origin data of
Korean and Japanese visitors to Fukuoka and Busan, this study cannot detect the
metro-to-metro impact of tourist flows between Fukuoka and Busan Metros. Based
on the proposed common platform for tourism research of Busan-Fukuoka
megapolitan cluster, a lot more detailed information can be collected and this will
help policymakers and planners to identify the potential tourism resources in the
development strategies.

Appendix

Table 9.14 Sector Classification of 2005 TIIO

10 Sector Classification		76 Sector Classification	
Code	Description	Code	Description
Intermediate sectors			
1	Agriculture, livestock, forestry, and fishery	001	Paddy
		002	Other grain
		003	Food crops
		004	Nonfood crops
		005	Livestock and poultry
		006	Forestry
		007	Fishery
2	Mining and quarrying	008	Crude petroleum and natural gas
		009	Iron ore
		010	Other metallic ore
		011	Nonmetallic ore and quarrying
3	Household consumption products(life-related manufacturing products)	012	Milled grain and flour
		013	Fish products
		014	Slaughtering, meat products, and dairy products
		015	Other food products
		016	Beverage
		017	Tobacco
		018	Spinning
		019	Weaving and dyeing
		020	Knitting
		021	Wearing apparel
		022	Other made-up textile products
		023	Leather and leather products
		024	Timber
		025	Wooden furniture
		026	Other wooden products
		060	Other manufacturing products

(continued)

10 Sector Classification		76 Sector Classification	
4	Basic industrial materials (primary makers' manufacturing products)	027	Pulp and paper
		028	Printing and publishing
		029	Synthetic resins and fiber
		030	Basic industrial chemicals
		031	Chemical fertilizers and pesticides
		032	Drugs and medicine
		033	Other chemical products
		034	Refined petroleum and its products
		035	Plastic products
		036	Tires and tubes
		037	Other rubber products
		038	Cement and cement products
		039	Glass and glass products
		040	Other nonmetallic mineral products
		041	Iron and steel
		042	Nonferrous metal
		043	Metal products
5	Processing and assembling (secondary makers' manufacturing products)	044	Boilers, engines, and turbines
		045	General machinery
		046	Metal-working machinery
		047	Specialized machinery
		048	Heavy electrical equipment
		049	Television sets, radios, audios, and communication equipment
		050	Electronic computing equipment
		051	Semiconductors and integrated circuits
		052	Other electronics and electronic products
		053	Household electrical equipment
		054	Lighting fixtures, batteries, wiring, and others
		055	Motor vehicles
		056	Motorcycles
		057	Shipbuilding
		058	Other transport equipment
		059	Precision machines
6	Electricity, gas and water supply	061	Electricity and gas
		062	Water supply
7	Construction	063	Building construction
		064	Other construction
8	Trade	065	Wholesale and retail trade
9	Transportation	066	Transportation

(continued)

10 Sector Classification		76 Sector Classification	
10	Services	067	Telephone and telecommunication
		068	Finance and insurance
		069	Real estate
		070	Education and research
		071	Medical and health service
		072	Restaurant
		073	Hotel
		074	Other services
		075	Public administration
		076	Unclassified

References

Andonian A, Kuwabara T, Yamakawa N, Ishida R (2016) The future of Japan's tourism: path for sustainable growth towards 2020. McKinsey Japan and Travel, Transport and Logistics Practice. https://www.mckinsey.com/~/media/mckinsey/industries/travel%20transport%20and%20logistics/our%20insights/can%20inbound%20tourism%20fuel%20japans%20economic%20growth/the%20future%20of%20japans%20tourism%20full%20report.ashx. Accessed 10 Jul 2019

Arai N (2011) Cross-strait tourism in the Japan–Korean border region: Fukuoka, Busan, and Tsushima. J Borderlands Stud 26(3):315–325

Bank of Korea Busan Office (2018) Trend analysis of Busan's struggling manufacturing activities and forecasts. Bank of Korea Press (written in Korean). https://www.bok.or.kr/ucms/cmmn/file/fileDown.do?menuNo=200570&atchFileId=KO_00000000000144080&fileSn=1. Accessed 15 Jul 2019

Benur AM, Bramwell B (2015) Tourism product development and product diversification in destinations. Tour Manag 50:213–224

Berry BJ (1980) Inner city futures: an American dilemma revisited. Trans Inst Br Geogr:1–28

Brouder P, Eriksson RH (2013) Tourism evolution: on the synergies of tourism studies and evolutionary economic geography. Ann Tour Res 43:370–389

Douglass M (2013) Decentralizing governance in a Transborder urban age: East Asia and the Busan-Fukuoka "common living sphere". Pac Aff 86(4):731–758

Fishman R (1987) The end of suburbia: a new kind of city is emerging-the "Technoburb". Los Angeles Times

Frenken K, Van Oort F, Verburg T (2007) Related variety, unrelated variety and regional economic growth. Reg Stud 41(5):685–697

Frey W (2003) Melting pot suburbs: a study of suburban diversity. In: Katz B, Lang R (eds) Redefining urban and suburban America: evidence from census 2000. Brookings Institution Press, Washington, DC, pp 155–180

Garlick S, Kresl P, Vaessen P (2006) The Øresund science region: a cross-border partnership between Denmark and Sweden. Organisation for Economic Co-operation

Hackworth J (2005) Emergent urban forms, or emergent post-modernisms? A comparison of large US metropolitan areas. Urban Geogr 26(6):484–519

Hall PG, Pain K (eds) (2006) The polycentric metropolis: learning from mega-city regions in Europe. Routledge, London

Hartman S (2016) Towards adaptive tourism areas? A complexity perspective to examine the conditions for adaptive capacity. J Sustain Tour 24(2):299–314

Henderson JC (2017) Destination development: trends in Japan's inbound tourism. Int J Tour Res 19(1):89–98

Kwon Y, Lim J, Kim E (2020) Diversifying visitor demand and its impact on Las Vegas's tourism industry during recovery from the Great Recession. Reg Sci Policy Pract 12(2):249–266

Lang RE, Knox PK (2009) The new metropolis: rethinking megalopolis. Reg Stud 43(6):789–802

Lang RE, LeFurgy JB (2007) Boomburbs: the rise of America's accidental cities. Brookings Institution Press, Washington, DC

Lang RE, Lim J, Danielsen KA (2020) The origin, evolution, and application of the megapolitan area concept. Int J Urban Sci 24(1):1–12

Martin R (2011) Regional economic resilience, hysteresis and recessionary shocks. J Econ Geogr 12(1):1–32

Meng B, Zhang Y, Inomata S (2013) Compilation and applications of IDE-JETRO's international input-output tables. Econ Syst Res 25(1):122–142

Miyazawa K (1976) Input-output analysis and the structure of income distribution. Springer-Verlag, New York, NY

Nelson A, Lang R (2011) Megapolitan America. Routledge, London

Sanz-Ibáñez C, Clavé SA (2014) The evolution of destinations: towards an evolutionary and relational economic geography approach. Tour Geogr 16(4):563–579

Sudjic D, Sayer P (1992) The 100 mile city. Harvest Books, Boston, MA

Takaki N, Lim JD (2011) Building an integrated trans-border economic region between Busan and Fukuoka. Seoul J Econ 24(2):197–220

Thyne M, Watkins L, Yoshida M (2018) Resident perceptions of tourism: the role of social distance. Int J Tour Res 20(2):256–266

Weidenfeld A (2013) Tourism and cross border regional innovation systems. Ann Tour Res 42:191–213

Chapter 10
Impact of Covid-19 in Tourism Regions. The Use of a Base Model for the Azores

Tomaz Ponce Dentinho

Abstract Confinement measures associated with Covid-19 have a major impact on the movement of people around the world and therefore on tourism. The aim of this chapter is to evaluate the impact of those measures in tourism regions looking at the Azores Island through an econometric model, inspired by the Economic Base Model. The model uses panel data, by year and island relative to basic regional activities: milk production, hosts, fisheries, public transferences and, negatively, regional public debt. Results allow to perceive the impact of Covid-19 on tourism and to estimate the effects in total employment and population per island. The conclusion is that islands with more tourism suffered a major impact but also recover faster.

Keywords Base model · Covid-19 · Tourism · Azores

10.1 Introduction

Several authors demonstrate that the population and regional employment are determined simultaneously (Greenwood 1985; Carlino and Mills 1987; Clark and Murphy 1996) and many subsequent works on the regional evolution of employment and population start from this evidence (Glaeser et al. 2001, 2004; Glaeser and Kohlhase 2003; Chi and Marcouiller 2013; Castro et al. 2020).

The population, or the supply side of work, results from the demographic evolution influenced by migrations, fertility and mortality, which are themselves marked by the context of the places (Waldorf and Franklin 2002). The demand for work is associated with the performance of basic activities or exports from local economies, explained by the Economic Base Model (Mulligan 2014). Migration

T. P. Dentinho (✉)
Universidade dos Açores, Angra do Heroísmo, Portugal
e-mail: tomas.lc.dentinho@uac.pt

© The Author(s), under exclusive licence to Springer Nature Singapore Pte Ltd. 2021
S. Suzuki et al. (eds.), *Tourism and Regional Science*, New Frontiers in Regional Science: Asian Perspectives 53, https://doi.org/10.1007/978-981-16-3623-3_10

results from the imbalance between the supply and demand for work (Isserman 1986; Plane and Rogerson 1994).

In small, open and island economies, where migration from and to outside communities is a current phenomenon, the key factor in population dynamics is the demand for work defined by basic activities, that is, those that are paid from abroad, and there may be a population that is unable to migrate and, if there is no labour market, remains self-sufficient with low-income levels, as is the case in many underdeveloped areas (Amaral et al. 2006). The employment estimates can be addressed through the estimation of econometric models based on the Economic Base Model (Fortuna and Vieira 2007; Dentinho and Fortuna 2018, 2019).

The Economic Base Model is a Keynesian-inspired model of real demand for the short term, which assumes that the performance of the regional economy depends on the variation of basic activities (Hoyt 1939; North 1955; Tiebout 1956; Krikelas 1992; Costa et al. 2009). These activities capture foreign resources (exports, public transfers, remittances, rents or other monetary inflows resulting from the establishment of equipments such as military bases, loans and, negatively, debt and interests payments). Prices are exogenous and there are no limitations from the supply side ensured by immigration in the event of a shortage of labour.

As mentioned by Galambos and Schreiber (1978), this model is a simple way to model a small economy, whose prices related to the ones in external markets. Consequently, it is possible to outline effective employment and settlement policies (Quintero 2007), taking into account that, in interconnected places, people migrate from places where there is unemployment to those where there is a demand for work. This type of model estimates the induced multiplier effects of an equivalent Input-Output model (Quintero 2007) and can be used to support regional growth policies at the local level or, in current terminology, place-based policies (Barca 2009; Duranton and Venables 2018).

This work estimates an econometric model, inspired by the Economic Base Model, with panel data, per year and per island, related to milk production, fishing, tourism, external public support and the incidence of Social Insertion Income (RSI). Data on public transfers and public debt to the Azores were also included, since these variables are not disaggregated by island, nor is there a proxy for this purpose. The employment of Portuguese people at the Lajes Air Base is also an explanatory variable. This model is an expansion of the econometric model of Dentinho and Fortuna (2018, 2019), but with data disaggregated by island. The model differs from the one elaborated by Fortuna and Vieira (2007) for the whole archipelago because we use variables more related to the basic activities.

The work structures as follows. Point 2 presents the employment model for the island. Point 3 includes a description of the data. The estimation of the model and its application for each island in the estimation of employment comes in points 4 and 5, respectively. In point 6, the population estimated by the Regional Statistics Service (SREA) is compared with the population estimated from the employment model. Based on the assumptions of the estimated model, point 7 shows the impacts on the population of market and policy scenarios for the period between 2020 and

2030, namely by considering the impacts of Covid-19 on tourism and on public debt. Finally, point 8 summarizes the main conclusions.

10.2   The Employment Model

According to the Economic Base Model, the regional economy depends on basic activities. In the Azores, the basic activities are agricultural and agro-industrial exports, strongly linked to the milk value chain, tourism, public transfers from abroad and fishing (Fortuna and Vieira 2007; Haddad et al. 2015) and, in a negative sense, the regional public debt of the previous year (Dentinho and Fortuna 2019). There was also added employment at Base das Lajes, justified by the work of Borba and Dentinho (2016), and the number of beneficiaries of the RSI, given their weight in some islands in the Azores.

The total employment per island, motivated by the objectives of the study, constitutes the dependent variable or to be explained. The exogenous, or explanatory, variables are milk production, used as a proxy for agricultural and agro-industrial exports; the number of guests in the hotel business, used as an indicator of tourist activity; public transfers before and after 2007—to capture the change in the Autonomous Regions' Finance Laws -,distributed across the islands according to the structure of expenditure on education; and the amount of public debt from the previous year, distributed by the islands according to population weight.

As in the study by Dentinho and Fortuna (2019), transfers from the European Union (EU) do not have a significant impact on the volume of employment, as does meat production, the coefficient of which is not statistically significant and has the opposite sign than expected. Finally, the consideration of health expenditure by island also does not have a statistically significant coefficient. For these reasons, these variables are not included in the selected model. However, some artificial variables were considered, which improve the adjustment of the model, for the islands of Graciosa, São Jorge, Faial, Flores and Corvo. Finally, using the employment obtained from the estimated model, the determination of the population is made with the aid of the relationship between population and employment.

10.3   The Data

Due to the lack of time series for all relevant variables, the data used to refer to the years 2003 to 2017 for eight islands and the years 2012 to 2017 for the island of Corvo. Since public employment data are only discriminated by island for the year 2012, we used the same structure to distribute regional public employment data by the island for the remaining years considered. We tried to distribute the employment data for the Azores based on the annual figures for the employment survey of the National Statistics Institute (INE), but the results obtained are lower than the ones

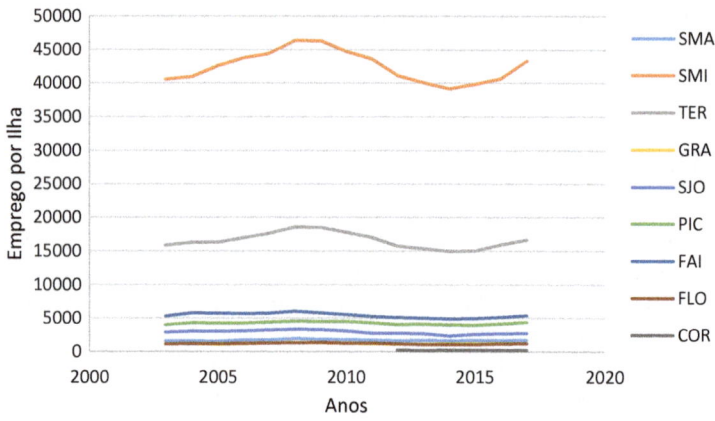

Fig. 10.1 Evolution of Employment in the Azores Islands

derived from the sum, for all the islands, of private and public employment reported by the statistics of Personnel Tables of private and public entities. Thus, the employment data we use corresponds to an activity rate of 30%, lower than the real activity rate of 40%. Nevertheless, if we assume that the relation between the employment reported by both sources stay the same we can work with the employment data discriminated by island.

Looking at Fig. 10.2a the liberalization of air transport in São Miguel, in 2014, and Terceira, in 2015, had a strong impact on the increase in the number of guests per island (Fig. 10.2a).

Figure 10.1 shows the evolution of employment by island. The impact of the crisis from 2008 to 2015 is evident, namely in the islands more integrated into foreign markets, such as São Miguel and Terceira, and relatively less dependent on public employment. A more detailed presentation of the employment series is made by comparing the actual data with the results obtained from the estimated model (Fig. 10.3).

Milk production respected the quota limits until 2008 when the reduction of milk production in the continent allowed to increase milk production in the Azores even before the end of the milk quota by countries in 2013. The island of São Miguel took better advantage of the liberalization of milk production, probably because there are more companies in the transformation and that compete for the purchase of the raw material (Fig. 10.2b).

The evolution of unloaded fishing, in tons, is strongly marked by the capture of tuna, which, in addition to having great annual variations and being concentrated in São Miguel and Pico, shows signs of decrease since 2010 (Fig. 10.2c).

State transfers show wide variations, decreasing sharply during the crisis. More recent data point to some recovery. On the other hand, public debt has been increasing significantly since 2008 (Fig. 10.2d). Given that transfers and debt work in the opposite direction in job creation, the effect of price developments, even if small, is cancelled and the model is clearer with values at current prices.

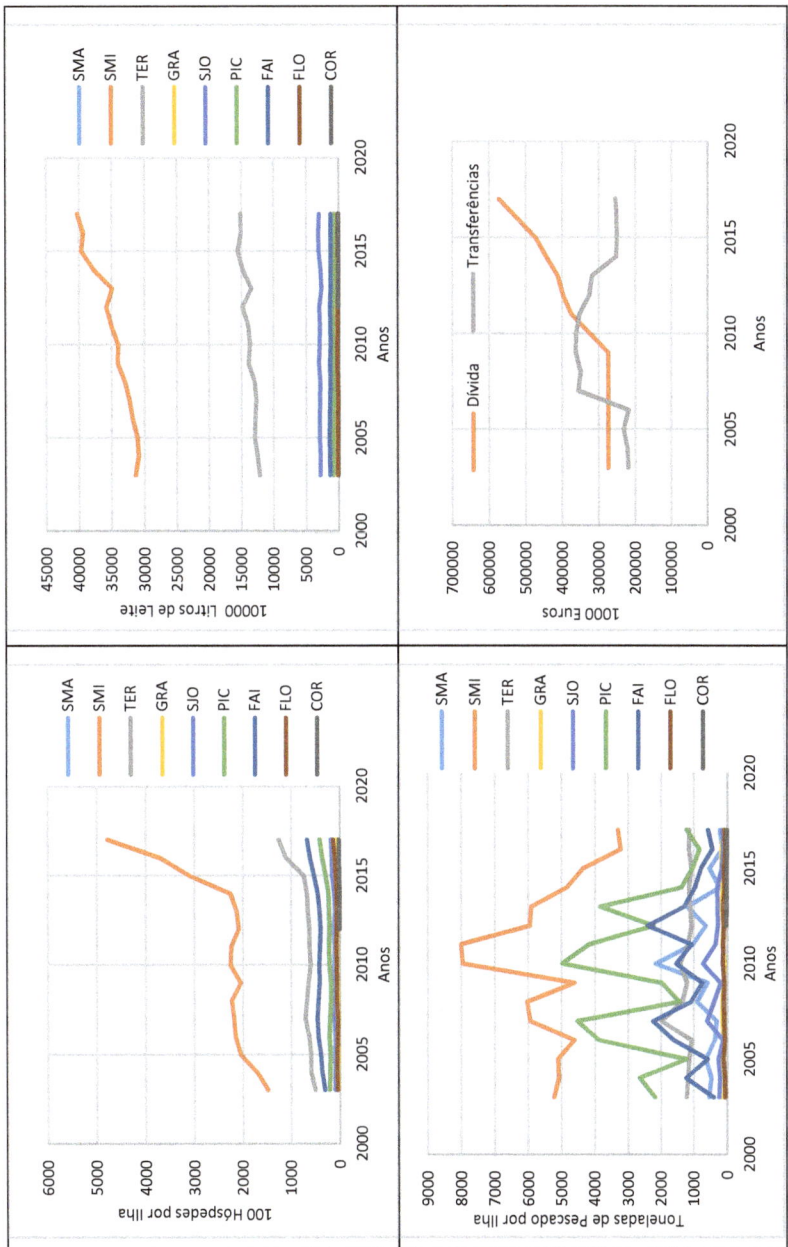

Fig. 10.2 Evolution of the Economic Base Factors of the Azores Islands. (**a**) 100 hosts per island; (**b**) 10,000 L of milk per island; (**c**) tons of fish captured per island; (**d**) thousands of Regional Public Debt and Public Transferences

Due to the specific characteristics of some islands, it was necessary to introduce artificial variables (dummy) for Graciosa, São Jorge, Faial, Flores and Corvo, so that the employment projections are acceptable for all islands in the archipelago. The coefficient of these variables is positive in the case of Faial and negative in the others. The inclusion of artificial variables for the islands of Pico and Santa Maria makes the fishing coefficient not significant, and they have not been considered.

10.4   Results

Table 10.1 includes the results of the estimation of the proposed model, and its analysis makes it possible to highlight some interesting aspects of the contribution of basic activities to the generation of employment.

Firstly, it should be noted that the model explains well the evolution of employment in each of the islands of the Azores, having as determinant factors the production of milk, the number of guests, the average fishing of 2 years, the transfers from the State and, as a negative factor, the public debt. We tried to use the debt service, but it was realized that this is not very stable over the years, due to the effect of amortizations.

Transfers from the EU showed little impact on employment per island, as did meat production, which, in addition to having a negative coefficient, is competitive with milk production. These variables were therefore not included in the final model.

Every 10,000 L of milk, corresponding to 1 hectare of pasture, generates 0.96 direct, indirect and induced jobs, which is similar to the value estimated by Dentinho

Table 10.1 Results of the Estimated Model for Employment

	Coefficients	Error	t	P
Constant	1597	154	10.38	0.000
Guests 100	3.37	0.26	12.82	0.000
Milk 10,000 L	0.96	0.08	12.53	0.000
Fisheries (tons)	0.20	0.07	2.80	0.006
State transfers (M. €)	136	9	15.85	0.000
State transfers after 2007 (M. €)	−44	3	−14.34	0.000
Previous year's debt (M €)	−60	6	−10.46	0.000
Social insertion income (number)	−0.45	0.13	−3.54	0.001
Portuguese employment at Lajes Base (number)	1.37	0.39	3.48	0.001
Graciosa Island (0 or 1)	−1216	185	−6.56	0.000
São Jorge Island (0 or 1)	−2029	251	−8.10	0.000
Faial Island (0 or 1)	1001	153	6.55	0.000
Flores Island (0 or 1)	−899	174	−5.16	0.000
Corvo Island (0.1)	−1414	235	−6.01	0.000
N. observations	126			
R^2 adjusted	0.99			

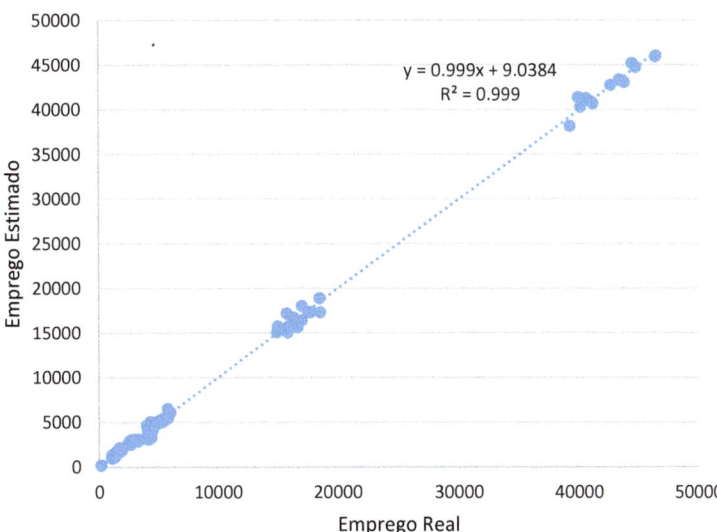

Fig. 10.3 Relationship between Real Employment and Estimated Employment

and Fortuna (2019) for the Azores as a whole. Every 100 additional guests generates 3.4 direct, indirect and induced jobs, which is close to the estimates presented by Dentinho and Fortuna (2019). Each ton of fish generates 0.20 direct, indirect and induced jobs.

Concerning the activity of the state, every million euros allocated generates 92 (= 136–44) direct, indirect and induced jobs. Each additional million of debt causes 60 direct, indirect and induced jobs loss. Each job at the Lajes Air Base generates 0.37 more jobs in addition to its own. Each person benefiting from RSI causes a loss of 0.45 jobs.

Figure 10.3 shows the relationship between real employment and estimated employment. The three groups of island dimensions (São Miguel, Terceira and the others) explain the three groups of points.

10.5   Real Employment and Estimated Employment by Island

The graphs in Fig. 10.4 illustrate the robustness of the estimated employment model, particularly for the larger islands (São Miguel and Terceira). Only the island of Pico seems to have deviated in recent years.

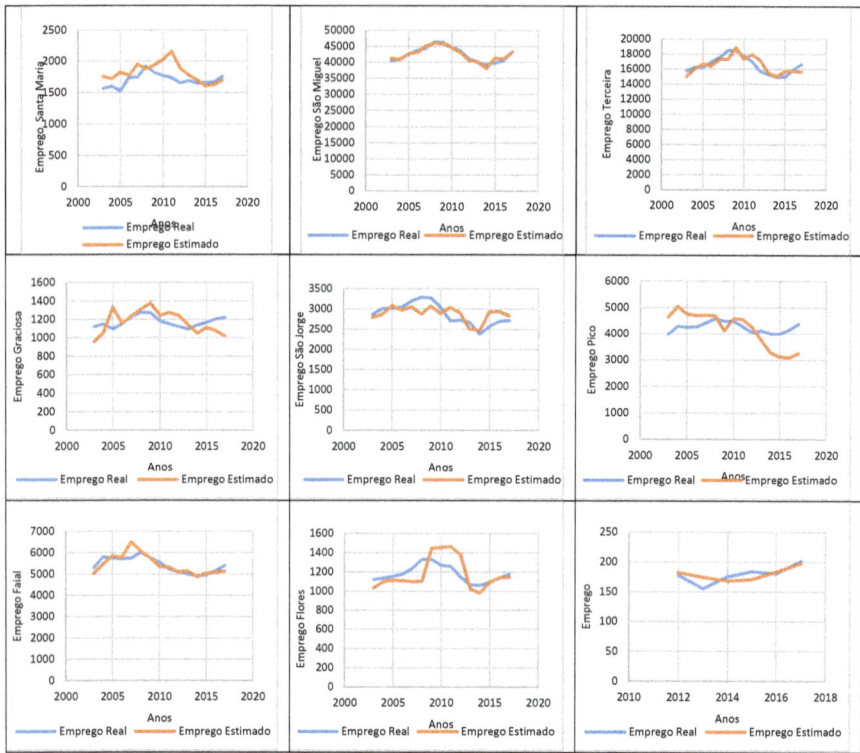

Fig. 10.4 Relationship between Real Employment and Estimated Employment by Azores Islands

10.6 Population and Population Estimated from the Employment Model

As we saw in point 1, the literature shows that regional population and employment are determined simultaneously. In the period under review, we have two population censuses (2001 and 2011) and employment data for 2001 and 2011. As there are nine islands, it is possible to relate the variation in employment between 2001 and 2011 with the variation in the population, annualizing the results. For the regression, an artificial variable was added for the islands of the old district capitals, which have a higher urban concentration (São Miguel, Terceira and Faial) and the relationship between employment variation and population variation was estimated. The results are shown in Table 10.2, where the dependent variable is the population variation.

The model results reveal two interesting phenomena. First, when the employment variation is nil, the population of the smallest islands decreases while, even so, the populations of the most urbanized islands increase (0.00665–0.00567), which indicates a process of common urban concentration to Portugal and the whole world. Real and Estimated Population in Fig. 10.5 reinforces the good results of model of Table 10.2.

Table 10.2 Results of the Population Variation Model given Employment Variation

	Coefficients	Error	t	P
Constant	−0.00567	0.00125	−4.52523	0.004
*Dummy*São Miguel, Terceira and Faial	0.00665	0.00174	3.81911	0.009
Employment variation	0.177637	0.07797	2.26211	0.064
N. observations = 9	9			
R^2	0.73			

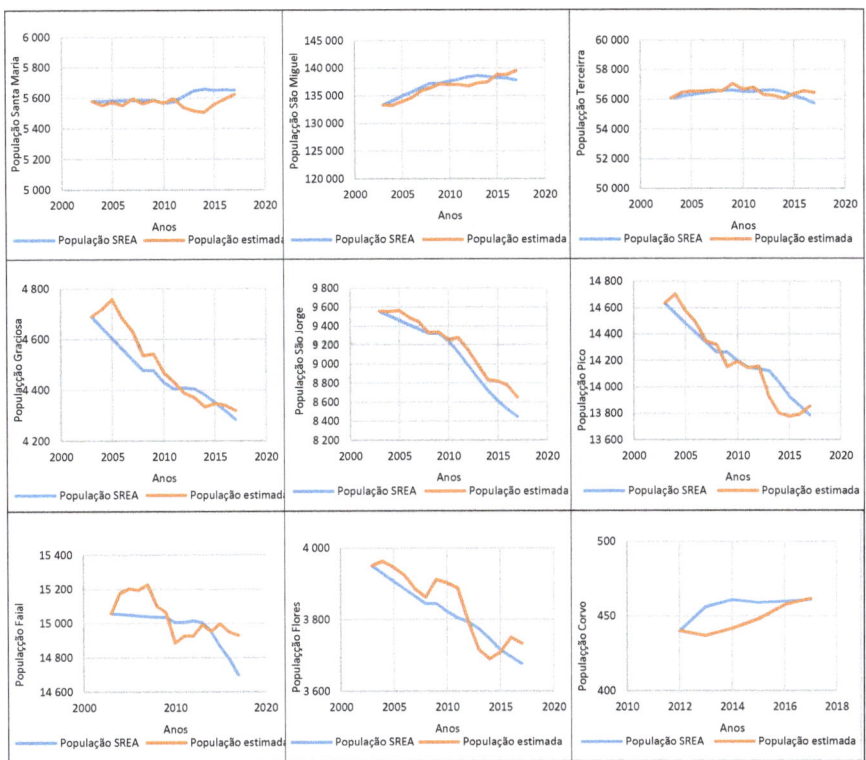

Fig. 10.5 Population of SREA and Population Estimated by the Model

Second, a percentage point of variation in employment translates into 0.18% points in the variation of the population, which is explained not only because the population is 3.33 higher than the employment considered in the model but also because there is a resilient population, about 40%, who do not migrate when they become unemployed, because they are poor and can survive in a self-subsistent way. As we can see in Fig. 10.5, only in the years of the crisis, SREA estimates contradict estimates based on employment models and, consequently, on the population, indicating that when the crisis is global people move from central places to where they migrate in Mainland Portugal, United States and Canada, back to the Azores where they can stay with lower costs of living.

10.7 Scenarios of the Impact of Tourism and Debt Associated with Covid-19

The model estimated and included in Table 10.1 makes it possible to estimate the impact of agriculture, tourism, fishing, public transfers and debt on the employment of each island. The model included in Table 10.2 allows estimating the evolution of the population with employment. Combining the two models, it is possible to make scenarios of population evolution given the panorama of evolution in agriculture, tourism, fishing, public transfers and debt in the population. The Covid-19 crisis is having a major impact on tourism around the world. This is also true for tourism in the Azores that, after a sharp increase associated with the liberalization of air transportation in São Miguel in 2014 and in Terceira in 2015, suffered a strong drop in 2020 due to Covid-19 (Fig. 10.6).

Looking at Fig. 10.6 the decrease in the number of hosts in 2020 is 70% for the whole of the Azores by higher for ore touristic places like São Miguel with a drop of

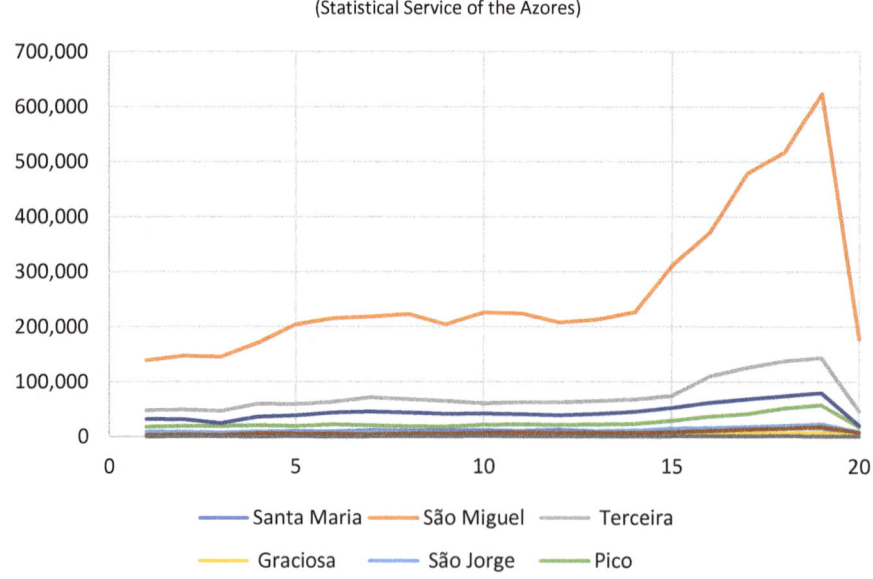

Fig. 10.6 Evolution of the Number of Hosts per Island of the Azores

Fig. 10.7 Population Evolution Scenarios with Debt of one billion Euros and Covid-19 Impact on Tourism

72%; interestingly the decrease was smaller for less infected islands like Flores (54%) and Corvo (26%).

The scenario that results from the maintenance of the public debt with value above two billion euros, allocated according to the distribution of the population, is very damaging to the economy of the islands. For the design of future scenarios we assume that it is possible to make agreed on a debt of one billion euros. On the exports, we assume that tourism grows at the rate of previous years in each of the islands, that milk is the average of the last 2 years, fishing is the average of the last 2 years and that transfers from the state grow at 1.22% per year. The market scenario admits the reduction of tourism in 2020 due to Covid-19. Results presented in Fig. 10.7 show a strong decrease in employment in 2020 and a slow increase after the Covid-19 crisis. Population will decrease in most islands but will recover faster in the more touristic islands of São Miguel, Pico, Faial and Flores.

194

T. P. Dentinho

10.8 Conclusion

This work includes a review of the literature on the economic and social determinants of the settlement of populations in a regional economy, with emphasis on the Economic Base Model. For the specific case of the Azores Island, milk production and transformation, tourism, fishing and transfers obtained under the Regional Finance Law have a positive effect on the generation of jobs and, consequently, on the fixation of the population in the islands. The islands with the highest urban concentration, however, tend to grow more in population for reasons other than the growth of regional basic employment. The employment generated by Base das Lajes also generates other jobs in the economy. The growth of public debt to high levels harms employment and the population. We estimate the impacts of Covid-19 on the employment and population of the island based on the data on hosts collected in 2020. It is concluded that tourism decrease had a major impact in the reduction of employment and, in the long term, population in the islands. Islands more dependent on tourism register a major drop but also recover better.

References

Amaral S, Vieira JC, Dentinho TP (2006) O impacto da Universidade do Huambo no desenvolvimento do planalto central de Angola. Revista Portuguesa de Estudos Regionais 13:5–28
Barca F (2009) An agenda for a reformed cohesion policy. A place-based approach to meeting European Union challenges and expectations. European Commission, Brussels
Borba JO, Dentinho TP (2016) Evaluation of urban scenarios using bid-rents of spatial interaction models as hedonic price estimators: an application to the Terceira Island, Azores. Ann Reg Sci 56(3):671–685
Carlino GA, Mills ES (1987) The determinants of county growth. J Reg Sci 27:39–54
Castro EA, Marques M, Marques JL, Viegas M (2020) Demografia e Economia: Um Modelo Regional Integrado de Estimação da População Portuguesa. Revista Portuguesa de Estudos Regionais 55:9–26
Chi G, Marcouiller DW (2013) Natural amenities and their effects on migration along the urban–rural continuum. Ann Reg Sci 50(3):861–883
Clark DE, Murphy CA (1996) Countywide employment and population growth: an analysis of the 1980s. J Reg Sci 36(2):235–256
Costa SJ, Delgado PA, Godinho MI (2009) "A teoria da base económica", Capítulo 14. In: Costa SJ, Peter Nijkamp P, Dentinho T (eds) Compêndio de Ciência Regional. Principia, Parede
Dentinho T, Fortuna M (2018) The impact of credit, public support and exports on regional growth. The case of the Azores. In: 12th world congress of the RSAI, 29 May–1 June, 2018, Goa, India
Dentinho T, Fortuna M (2019) How regional governance constrains regional development. Evidences from an Econometric Base model for the Azores. Revista Portuguesa de Estudos Regionais 52:25–35
Duranton G, Venables A (2018) Place-based policies for development. In: NBER Working Paper No. 24562
Fortuna M, Vieira JC (2007) The contribution of tourism to growth: lessons from the Azores and Madeira. Revista Turismo e Desenvolvimento 7(8):43–55

Galambos E, Schreiber A (1978) Making sense out of dollars: economic analysis for local government. In: WDCNLo (ed) Cities. National League of Cities, Washington, DC

Glaeser E, Kohlhase JE (2003) Cities, regions and the decline of transport costs. Rev Econ Des 83 (1):197–228

Glaeser EL, Kolko J, Saiz A (2001) Consumer City. J Econ Geogr 1(1):27–50

Glaeser EL, Saiz A, Burtless G, Strange WC (2004) The rise of the skilled city. In: Brookings-Wharton papers on urban affairs, pp 47–105

Greenwood MJ (1985) Human migration: theory, models, and empirical studies. J Reg Sci 20:521–544

Haddad E, Silva V, Porsse A, Dentinho T (2015) Multipliers in an island economy: the case of the Azores. In: The region and trade

Hoyt H (1939) The structure and growth of residential neighborhoods in American cities. U.S. Government Printing Office, Washington, DC

Isserman A (ed) (1986) Population change and the economy. Kluwer, Boston

Krikelas AC (1992) Why regions grow: a review of research on the Economic Base model. Fed Reserve Bank Atlanta Econ Rev 77:16–29

Mulligan G (2014) Regional science at sixty: traditional topics and new directions. Australas J Reg Stud 20(1):20–34

North DC (1955) Location theory and regional economic growth. J Polit Econ 63:243–258

Plane D, Rogerson P (1994) The geographical analysis of population. Wiley, New York

Quintero JP (2007) Regional economic development: an Economic Base study and shift-share analysis of Hays County, Texas. Applied Research Projects: Texas State Univerity Public Administration Program

Tiebout CM (1956) A pure theory of local public expenditures. J Polit Econ 64:416–424

Waldorf B, Franklin R (2002) Spatial dimensions of the Easterlin hypothesis fertility variations in Italy. J Reg Sci 42:549–578